Lecture Notes in Artificial Intelligence 7993

Subseries of Lecture Notes in Computer Science

T0240527

Tetsuo Ida   Jacques Fleuriot (Eds.)

Automated Deduction in Geometry

9th International Workshop, ADG 2012
Edinburgh, UK, September 17-19, 2012
Revised Selected Papers

 Springer

Volume Editors

Tetsuo Ida
University of Tsukuba
Department of Computer Science
Tsukuba 305-8573, Japan
E-mail: ida@cs.tsukuba.ac.jp

Jacques Fleuriot
The University of Edinburgh
School of Informatics
Informatics Forum
10 Crichton Street
Edinburgh EH8 9AB, UK
E-mail: jdf@inf.ed.ac.uk

ISSN 0302-9743   e-ISSN 1611-3349
ISBN 978-3-642-40671-3   e-ISBN 978-3-642-40672-0
DOI 10.1007/978-3-642-40672-0
Springer Heidelberg New York Dordrecht London

Library of Congress Control Number: 2013948305

CR Subject Classification (1998): I.2.3, I.3.5, F.4.1, F.3, G.2-3, D.2.4

LNCS Sublibrary: SL 7 – Artificial Intelligence

© Springer-Verlag Berlin Heidelberg 2013

Typesetting: Camera-ready by author, data conversion by Scientific Publishing Services, Chennai, India

Printed on acid-free paper

Springer is part of Springer Science+Business Media (www.springer.com)

Preface

Automated Deduction in Geometry (ADG) is a well-established international series of workshops that bring together researchers interested in the theoretical and practical aspects of machine-based geometric reasoning. Since its inception in 1992 – initially as a workshop on purely algebraic approaches for geometric reasoning – ADG has continually evolved, enabling it to add a diversity of new topics and applications to its scope and establish its status as the foremost biennial event for automated reasoning in geometry.

In 2012, the ADG workshop was held in the School of Informatics at the University of Edinburgh, UK, during September 17–19. Each of the 15 submissions accepted for presentation was reviewed by at least two members of the Program Committee chaired by T. Ida, resulting in a lively 3-day program that also included invited talks by M. Beeson from San Jose State University, USA, and D. Wang from Université Pierre et Marie Curie - CNRS, France. As is customary, the workshop was followed by a new, open call-for-papers, and the present proceedings, with its 10 accepted articles detailing work at the forefront of research in automated deduction in geometry, is the result of a fresh refereeing and paper-revision process. We briefly review the various contributions next, with the understanding that our broad categorization is not meant to be exhaustive or absolute.

The first group of papers is mainly concerned with advances in and applications of algebraic techniques: J. Yang, D. Wang, and H. Hoong continue their investigation of reparameterization of plane curves and present a method that uses C^1 piecewise Möbius transformation in an effort to improve angular speed uniformity. S. Moritsugu presents an extension of Descartes Circle Theorem to Steiner's porism involving n circles and, with the help of Gröbner bases and resultant methods, produces several novel results along the way. P. Mathis and P. Schreck discuss an application of Cayley-Menger determinants to yield systems of equations that can be used for geometric constraints involving points and hyperplanes. C. Borcea and I. Streinu address issues in a generalized form of rigidity theory by considering volume frameworks on hypergraphs. J. Bowers and I. Streinu investigate computational origami using Lang's Universal Molecule algorithm and present several new results about the 3D foldability of origami. F. Ghourabi, A. Kasem, and C. Kaliszyk provide a rigorous reformulation of Huzita's axioms for fold operations in computational origami and give a single generalized fold operation that replaces all the reformulated operations.

The second group of papers broadly looks at geometric reasoning from a logical and theorem proving standpoint: C. Brun, J.-F. Dufourd, and N. Magaud use the Coq proof assistant to mechanize an incremental procedure for determining the convex hull of a set of points and discuss the derivation by hand of its associated imperative C++ program. G. Braun and J. Narboux carry out a study in the mechanization of Hilbert's and Tarski's axiomatics in Coq and show how Hilbert's axioms can be derived from Tarski's. U. Siddique, V. Avrantinos, and S. Tahar explore the formalization of geometrical optics in the proof assistant HOL Light and use their new mechanical framework to verify the stability of two widely-used optical resonators. S. Stojanovic proposes two heuristics for improving the efficiency of forward chaining in the coherent logic theorem prover ArgoCLP and shows that these improve automated proofs in a Hilbert-like geometry.

On the invited papers front, we include the abstract of D. Wang's ADG 2012 invited talk, which covered historical aspects of the mechanization of geometry and discussed the engineering foundations needed to develop the next generation of geometric reasoning software. In a happy marriage between algebra and logic, the article by M. Beeson discusses the relationships between proof and computation in geometry from both theoretical and practical standpoints. He introduces a new vector geometry theory and reports on experiments that solve several automated reasoning challenges in Tarski's geometry set by Quaife in 1990.

We thank the School of Informatics for hosting ADG 2012 and for providing generous financial support. Our gratitude also extends to Suzanne Perry for her valuable help in organizing the workshop and its associated events. Last, but by no means least, we thank the Program Committee and secondary referees for their reviews and discussions – both for the workshop papers and the current proceedings. Their dedication is what makes ADG a continuing success.

May 2013   Tetsuo Ida
                                                         Jacques Fleuriot

Organization

Program Committee

Hirokazu Anai	Fujitsu Laboratories Ltd., Japan
Francisco Botana	Universidad de Vigo, Spain
Xiaoyu Chen	Beihang University, China
Giorgio Dalzotto	Università di Pisa, Italy
Jacques Fleuriot	University of Edinburgh, UK
Laureano Gonzalez-Vega	Universidad de Cantabria, Spain
Hoon Hong	North Carolina State University, USA
Tetsuo Ida	University of Tsukuba, Japan
Andres Iglesias Prieto	Universidad de Cantabria, Spain
Predrag Janicic	University of Belgrade, Serbia
Deepak Kapur	University of New Mexico, USA
Ulrich Kortenkamp	Martin-Luther-Universität Halle-Wittenberg, Germany
Shuichi Moritsugu	University of Tsukuba, Japan
Julien Narboux	Université de Strasbourg, France
Pavel Pech	University of South Bohemia, Czech Republic
Tomas Recio	Universidad de Cantabria, Spain
Georg Regensburger	Johann Radon Institute for Computational and Applied Mathematics, Austria
Jürgen Richter-Gebert	Technische Universität München, Germany
Pascal Schreck	Université de Strasbourg, France
Meera Sitharam	University of Florida, USA
Thomas Sturm	Max-Planck-Institut für Informatik, Saarbrücken, Germany

Additional Reviewers

Montes Antonio
Robert Joan Arinyo
Filip Maric
Uwe Waldmann
Christoph Zengler

Table of Contents

Proof and Computation in Geometry

Michael Beeson

San José State University, San José, CA, USA

Abstract. We consider the relationships between algebra, geometry, computation, and proof. Computers have been used to verify geometrical facts by reducing them to algebraic computations. But this does not produce computer-checkable first-order proofs in geometry. We might try to produce such proofs directly, or we might try to develop a "back-translation" from algebra to geometry, following Descartes but with computer in hand. This paper discusses the relations between the two approaches, the attempts that have been made, and the obstacles remaining. On the theoretical side we give a new first-order theory of "vector geometry", suitable for formalizing geometry and algebra and the relations between them. On the practical side we report on some experiments in automated deduction in these areas.

1 Introduction

The following diagram should commute:

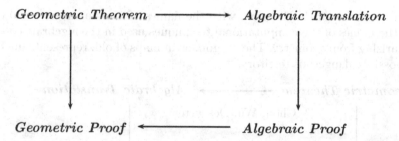

That diagram corresponds to the title of this paper, in the sense that proof is on the left side, computation on the right. The computations are related to geometry by the two interpretations at the top and bottom of the diagram. In the past, much work has been expended on each of the four sides of the diagram, both in the era of computer programs and in the preceding centuries. Yet, we still do not have machine-found or even machine-checkable geometric proofs of the theorems in Euclid Book I, from a suitable set of first-order axioms–let alone the more complicated theorems that have been verified by computerized algebraic computations.[1] In other words, we are doing better on the right side of the diagram than we are on the left.

---

[1] A very good piece of work towards formalizing Euclid is [1], but because it mixes computations (decision procedures) with first-order proofs, it does not furnish a counterexample to the statement in the text.

T. Ida and J. Fleuriot (Eds.): ADG 2012, LNAI 7993, pp. 1–30, 2013.

First-order geometrical proofs are beautiful in their own right, and they give more information than algebraic computations, which only tell us that a result is true, but not why it is true (i.e. what axioms are needed and how it follows from those axioms). Moreover there are some geometrical theorems that cannot be treated algebraically at all (because their algebraic form involves inequalities).

We will discuss the possible approaches to getting first-order geometrical proofs, the obstacles to those approaches, and some recent efforts. In particular we discuss efforts to use a theorem-prover or proof-checker to facilitate a "back translation" from algebra to geometry (along the bottom of the diagram). This possibility has existed since Descartes defined multiplication and square root geometrically, but has yet to be exploited in the computer age. According to Chou *et. al.* ([8], pp. 59–60), "no single theorem has been proved in this way."

To accomplish that ultimate goal, we must first bootstrap down the left side of the diagram as far as the definitions of multiplication and square root, as that is needed to interpret the algebraic operations geometrically. We will discuss the progress of an attempt to do that, using the axiom system of Tarski and resolution theorem-proving.

1.1   That Commutative Diagram, in Practice

In theory, there is no difference between theory and practice.
In practice, there is.

– Yogi Berra

Here is a version of the diagram, with the names of some pioneers[2], and on the right the names of the computational techniques used in the algebraic computations arising from geometry. The dragon, as in maps of old, represents uncharted and possibly dangerous territory.

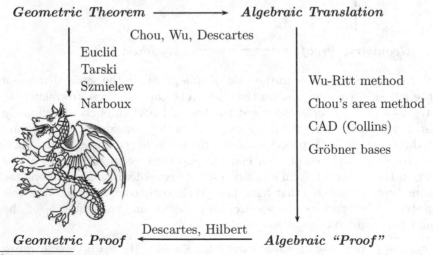

Geometric Theorem ⟶ *Algebraic Translation*

Chou, Wu, Descartes

Euclid
Tarski
Szmielew
Narboux

Wu-Ritt method

Chou's area method

CAD (Collins)

Gröbner bases

Descartes, Hilbert

Geometric Proof ⟵ *Algebraic "Proof"*

---

[2] Many others have contributed to this subject, including Gelernter, Gupta, Kapur, Ko, Kutzler and Stifter, and Schwabhäuser.

Our proposal is the following:

(i) The goal is the lower left, i.e. first-order geometric proofs.

(ii) Finding them directly is difficult (hence the dragon in the picture).

(iii) Therefore: let's get around the dragons by going across the bottom from right to left.

1.2   Issues Raised by This Approach

The first issue is the selection of a language and axioms (that is, a theory) in which to represent geometrical theorems and find proofs. We discuss that briefly in the next section, and settle upon Tarski's language and the axioms for ruler-and-compass geometry.

The next issue is this: if we want to go around the right side of the diagram and back across the bottom, and end up with a proof, then the computational part (the algebra on the right side of the diagram) will have to be formalized in some theory. In other words, we will have to convert computations to *verified*, or *formal* computations (proofs in some algebraic theory).[3] A formal theory of algebra will be required.

The third issue is, how can we connect the left and right sides of the diagram? If we have a formal geometrical theory on the left, and a formal algebraic theory on the right, we need (at the least) two formal translation algorithms, one in each direction. Technically such mappings (taking formulas to formulas) are called "interpretations"; we will need them to take proofs to proofs as well as formulas to formulas.

That approach promises to be cumbersome: two different formal theories, two formal interpretations, algebraic computations, and proofs verifying the correctness of those computations. We will cut through some of these complications by exhibiting a new formal theory **VG** of "vector geometry." This theory suffices to formalize the entire commutative diagram, i.e. both algebra and geometry. The first half of this paper is devoted to the formal theories for geometry, algebra, and vector geometry, and some metatheorems about those theories.

Within that theoretical framework, there is room for a great deal of practical experimentation. We have carried out some preliminary experiments, on which we report. In these experiments, we used the resolution-based theorem provers Otter and Prover9, but that is an arbitrary choice; one could produce proofs by hand using Coq as in [20][4] or in another proof-checker, or using another theorem-prover.

---

[3] This is related to the general problem of verifying algebraic computations carried out by computer algebra systems, which often introduce extra assumptions that occasionally result in incorrect results.

[4] There is an issue about how easy it is or is not to extract first-order geometric proofs from a Coq proof. In my opinion it should be possible, but Coq proofs are not *prima facie* first-order.

2   First Order Theories of Geometry

In this section we discuss the axiomatization of geometry, and its formalization in first-order logic. These are not quite the same thing, as there is a long history of second order axiomatizations (involving sets of points). Axiomatizations have been given by Veblen [33], Pieri [24], Hilbert [14], Tarski [31], Borsuk and Szmielew [5], and Szmielew [29], and that list is by no means comprehensive.

The following issues arise in the axiomatization of geometry:

- What are the primitive sorts of the theory?
- What are the primitive relations?
- What (if any) are the function symbols?
- What are the continuity axioms?
- How is congruence of angles defined?
- How is the SAS principle built into the axioms?
- How close are the axioms to Euclid?
- Are the axioms few and elegant, or numerous and powerful?
- Are the axioms strictly first-order?
- Can the axioms be stated in terms of the primitives, or do they involve defined concepts?
- Do the axioms have a simple logical form (e.g. universal or $\forall\exists$)?

Evidently there is no space to discuss even the few axiomatizations mentioned above with respect to each of these issues; we point out that the answers to these questions are more or less independent, which gives us at least $n = 2^{11}$ different ways to formalize geometry, whose relationships and mutual interpretations can be studied. Nearly every possible combination of answers to the "issues" has something to recommend it. For example, Hilbert has several sorts, and his axioms are not strictly first-order; Tarski has only one sort (points) and ten axioms. My own theory of constructive geometry [4,3] has points, lines, and circles, and function symbols so that the axioms are quantifier-free and disjunction-free.

In Euclid, geometry involves lines, line segments, circles, arcs, rays, angles, and "figures" (polygons). Rays and segments are needed only for visual effect, so a formal theory can dispense with them.

Hilbert [14] treated angles as primitive objects and angle congruence as a primitive relation. But angles can be treated as ordered triples of points, so they too can be dispensed with, as we will now show.[5] The key idea is Tarski's "five-segment axiom" (A5), shown in Fig. 1.

If the four solid segments in Fig. 1 are pairwise congruent, then the fifth (dotted) segments are congruent too. This is essentially SAS for triangles dbc and DBC. The triangles abd and ADE are surrogates, used to express the congruence of angles dbe and DBE. By using Axiom A5, we can avoid all mention of angles.

In fact, we don't even need lines and circles; every theorem comes down to constructing some points from given points, so that the constructed points bear

---

[5] The idea to define these notions (instead of take them as primitive) goes back (at least) to J. Mollerup [19], but he attributes it to Veronese.

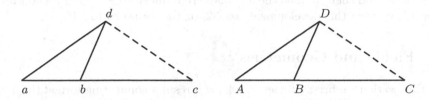

Fig. 1. The five-segment axiom (A5)

certain relations to the original points. Realizing this, Tarski formulated (in 1926) his theories of geometry using only one sort of variables, for points.

The fundamental relations to be mentioned in geometry are usually (at least for the past 120 years) taken to be *betweenness* and *equidistance*. We write $B(a, b, c)$ for "a, b, and c are collinear, and b is strictly between a and c." Similarly $T(a, b, c)$ for non-strict betweenness: either $B(a, b, c)$ or $a = b$ or $b = c$. T stands for "Tarski"; Hilbert used strict betweenness.[6] Equidistance is formally written $E(a, b, c, d)$, but often written informally as $ab \equiv cd$ or $ab = cd$.

The question of function symbols is related to the issue of logical form. For example, we may wish to introduce $ext(a,b,c,d)$ to stand for the point extending segment ab past b by the amount cd. In this way, we can reduce a $\forall\exists$ axiomatization to a universal one.

The question of continuity axioms brings us to the distinction between second order and first-order axiomatizations. Tarski seems to have been the first to give a first-order account of Euclidean geometry by restricting his continuity schema to first-order instances. His paper [31] is called *What is Elementary Geometry*, and he called his first-order theory "elementary geometry" to emphasize its first-order nature. We call this theory "Tarksi geometry". Because "elementary" means first-order, the word is not available for what is usually now known as "ruler and compass geometry."[7] In ruler and compass geometry, the infinite schema of all first-order continuity axioms is replaced by two consequences, "line-circle continuity" and "circle-circle continuity."

In the rest of this paper, we will work with Tarski's axioms for ruler and compass geometry. These axioms are known as A1 through A10, plus the line-circle and circle-circle continuity axioms. We chose to work with this theory because among geometrical theories, it is the simplest in the sense of having only one sort of variables, two primitive relations, and a small number of axioms that do not need defined notions to express. Starting with such pristine axioms requires a long development to reach the level of Euclid; this formal development was carried out by Tarski in his 1956-57 lecture notes, starting from a larger set

---

[6] Betweenness, which does not occur in Euclid, was introduced by Moritz Pasch [21]; see also [16] for another early paper on betweenness.

[7] That would have otherwise been natural, since Euclid's work is titled the *Elements*. We can't call ruler and compass geometry "Euclidean", either, since that has come to mean geometry with the parallel axiom, as opposed to "non-Euclidean geometry."

of axioms; and then by 1965 the axioms were reduced to A1-A10 plus continuity. For a history of this development see [32] or the foreword to [29].

3   Fields and Geometries

In this section we briefly review the known results connecting formal theories of geometry with corresponding formal theories of algebra (field theory).

A Euclidean field is an ordered field in which every positive element has a square root; or equivalently (without mentioning the ordering), a field in which every element is a square or minus a square, every element of the form $1 + x^2$ is a square, and -1 is not a square.

If $\mathbb{F}$ is a Euclidean field, then using analytic geometry we can expand $\mathbb{F}^2$ to a model of ruler and compass geometry.

Descartes and Hilbert showed, by giving geometric definitions of addition, multiplication, and square root, that every model of Euclidean geometry is of the form $\mathbb{F}^2$, where $\mathbb{F}$ is a Euclidean field.

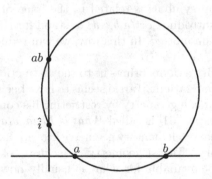

Fig. 2. Multiplication according to Hilbert

Similarly every model of Tarski geometry is $\mathbb{F}^2$, where $\mathbb{F}$ is real-closed. The smallest model of Tarski geometry corresponds to the case when $\mathbb{F}$ is the field of real algebraic numbers. The smallest model of ruler and compass geometry is the "Tarski field" $\mathbb{T}$, defined as the least subfield of the reals closed under square roots of positive elements. In a natural sense, $\mathbb{T}^2$ is the minimal model of ruler and compass geometry EG. $\mathbb{T}$ consists of all real algebraic numbers whose degree over $\mathbb{Q}$ is a power of 2.

3.1   Models and Interpretations

In general model-theoretic arguments are looked at by proof theorists as "interpretations." An interpretation maps formulas ϕ of the source theory into formulas $\hat{\phi}$ of the target theory, preserving provability:

$$\vdash \phi \Rightarrow \vdash \hat{\phi}$$

Usually the proof also shows how to transform the proofs efficiently. Generally interpretations have several advantages over models, all stemming from their greater explicitness. The main advantage is that an interpretation enables one to translate proofs from one theory to another. A model theoretic theorem, coupled with the completeness theorem, may imply the existence of a proof, but not give the slightest clue how to find it; while interpretations often give a linear-time proof translation.

There is a price to be paid, as the technical details of interpretations are often more intimidating than those of the corresponding model-theoretic arguments. Nevertheless, if we hope to use the equivalences of geometry and algebra to find proofs, model theory will not suffice. We need explicit interpretations.

Another reason for working with interpretations is that they can also be used for theories with non-classical logics, for example intuitionistic logic. The details of the interpretations between geometry and field theory can be found in [3]. Below, after introducing a theory of vector geometry, we will give a sample of these details.

3.2   Interpretations between Geometry and Algebra

Here we sketch the main ideas connecting formal theories of geometry and algebra, indicating how these ideas can be expressed using interpretations rather than model theory. Since at this point we have not given an explicit list of axioms, the discussion cannot be completely precise, but still the ideas can be explained. When we go from algebra to geometry, we fix a line L, which we call the "x-axis", containing two fixed distinct points α and β. (These interpret the scalars 0 and 1.) Then we show that one can construct a line perpendicular to any given line K, passing through a given point q, without needing a case distinction as to whether q is or is not on K. Since we do not have variables for lines, we need two points (say k_1 and k_2) to specify K and two to specify the resulting line, so we need two terms $perp_1(k_1, k_2, q)$ and $perp_2(k_1, k_2, q)$ that determine this perpendicular. Then we define "the y-axis" to be the perpendicular to the x-axis at α. We can then define coordinate functions X and Y by terms of our geometrical theory, such that for any point q, $X(q)$ is a point P on the fixed line L such that the line containing $X(q)$ and q is perpendicular to L, and the line containing q perpendicular to the y-axis meets the y-axis at a point Q such that $Q\alpha \equiv \alpha Y(q)$. (Notice that Q is on the y-axis but $Y(q)$ is on the x-axis.) It is not at all trivial to construct these terms without a test-for-equality function (symbol), but it can be done (see [3]).[8] Then we can find a term F of our geometrical theory that takes two points x and y on the x-axis and constructs

---

[8] This permits us to eschew a test-for-equality symbol, which is good, for two reasons: nothing like a test-for-equality construction occurs in Euclid, and simpler is better. But [3] uses terms for the intersections of lines and circles; whether those can be eliminated is not known.

the point p whose coordinates are x and y. One first constructs the point Q on the y-axis such that $\alpha Q \equiv \alpha y$, and the the lines perpendicular to the x axis at x and perpendicular to the y axis at Q. One needs the parallel postulate (A10) to prove that these lines actually meet in the desired point $F(x, y)$.

Part of the price to pay for using interpretations instead of models is that the algebraic interpretation ϕ^* of a geometric formula ϕ has two free variables for each free variable of ϕ, one for each coordinate. Then when we translate back into geometry, these two variables do not recombine, but become two different point variables, restricted to the x-axis. They must be recombined using F.

3.3   Euclid Lies in the AE Fragment

By the AE fragment, or the $\forall\exists$ fragment, we mean the set of formulas of the form $\forall x \exists y\, A(x, y)$, where x and y may stand for zero or more variables, and A is quantifier-free.

Euclid's theorems have the form,

> *Given some points bearing certain relations to each other, there exist (one can construct) certain other points bearing specified relations to the original points and to each other.*

The case where no additional points are constructed is allowed. The points are to be constructed with ruler and compass, by constructing a series of auxiliary points. Constructed points are built up from the intersections of lines and circles.

Theorems of this form can be translated into Euclidean field theory (formulated with a function symbol for square root). Since the intersections of lines and circles, and the intersections of circles, can be expressed using only quadratic equations, and there is a function symbol for square root, constructed points correspond to terms of the theory. Euclid's theorems are thus in $\forall\exists$ form, both before and after translation into algebra.

A careful analysis of Euclid's proofs shows that, apart from some case distinctions as to whether two points are equal or not, or a point lies on a line or not, the proofs are constructive: Euclid provides a finite number of terms, one of which works in each case. This is closely related to Herbrand's theorem, which would tell us that if $\forall x \exists y\, A(x, y)$ is provable, then there are finitely many terms $t_1, \ldots, t_n$ such that the disjunction of the $A(x, t_i(x))$ is provable.

Some parts of Euclid are about "figures", which are essentially arbitrary polygons. Euclid did not have the language to express these theorems precisely, since that would require variables for finite sets or lists or points, but we regard them as "theorem schemata", i.e. for each fixed number of vertices, we have a Euclidean theorem.

The AE fragment of the modern first-order theory of ruler and compass geometry is thus the closest thing we have to a formal analysis of Euclid. Euclid did not study theorems with more alternations of quantifiers.

3.4   Decidability Issues

A proper study of the relations between proof and computation in geometry must take place against the backdrop of the many known, and a few unknown, results about the decidability or undecidability of various theories. After all, the decidability of a formal theory means that provability in the theory can be reduced to computation. We offer in this section a summary of these known and unknown results.

Gödel and Church showed that in number theory, provability cannot be reduced to computation; Tarski showed that in geometry, it can, in the sense that Tarski geometry can be reduced to the theory RCF of real closed fields, for which Tarksi gave a decision procedure. Later Fischer and Rabin [10] showed that any decision procedure for RCF is at least exponential in the length of the input, and others showed it is at least double exponential in the number of variables; and Collins gave a decision procedure that is no worse than that bound (Tarski's was). (See [27] for more details.) Thus from a practical point of view it doesn't do us any good to know that RCF is decidable. There are interesting questions that can be formulated in RCF, questions whose answers we do not know, but if they involve more than six variables, then we are not going to compute the answers by a decision procedure for RCF.

On the other hand, if we drop the continuity axioms entirely, we get back the complications of number theory. Julia Robinson [28] proved that $\mathbb{Q}$ is an undecidable field, and later extended this result to algebraic number fields. Regarding the decidability of theories rather than particular fields, Ziegler [36] proved that any finitely axiomatizable extension of field theory is undecidable–in particular the theory of Euclidean fields. His proof shows the AEA fragment is undecidable. (Here AEA means $\forall\exists\forall$, formulas with three blocks of unlike quantifiers as indicated.) It does not say anything about the AE fragment. It is presently an open problem whether the AE fragment of RCF (and hence the AE fragment of Tarski geometry) is decidable. The fact that Euclid's *Elements* lies within this fragment focuses attention on the problem of its decidability.

Tarski conjectured that $\mathbb{T}$ (recall that $\mathbb{T}^2$ is the smallest model of rule and compass geometry) is undecidable, but this is still an open problem. Since $\mathbb{T}$ is not of finite degree over the rationals, its undecidability is not implied by Julia Robinson's results about algebraic number fields.

4   Tarski's Ruler and Compass Geometry

In this section we comment on the axioms of Tarski's theory, which can be found in full formal detail in [32] or [29]. This section is intended both as an introduction to Tarski's axioms, and as a description of the Skolem symbols we added to make the theory quantifier-free instead of $\forall\exists$. As mentioned above, the primitives are non-strict betweenness T and segment congruence $ab \equiv cd$, which is a 4-ary relation between points.

4.1   Tarski's First Six Axioms

Axiom A5 has been discussed and illustrated above. The other five are

$$uv \equiv vu \qquad \text{(A1)}$$
$$uv \equiv wx \wedge uv \equiv yz \rightarrow wx \equiv yz \qquad \text{(A2)}$$
$$uv \equiv ww \rightarrow u = v \qquad \text{(A3)}$$
$$T(u, v, ext(u, v, w, x)) \qquad \text{(A4), segment extension}$$
$$T(u, v, u) \rightarrow u = v \qquad \text{(A6)}$$

We have added a Skolem symbol to express (A4) without a quantifier.

4.2   Pasch's Axiom (1882)

Moritz Pasch [21] (See also [22], with an historical appendix by Max Dehn)
supplied an axiom that repaired many of the defects that nineteenth-century
rigor found in Euclid. Roughly, a line that enters a triangle must exit that
triangle. As Pasch formulated it, it is not in AE form. There are two AE versions,
illustrated in Fig. 3. These formulations of Pasch's axiom go back to Veblen [33],
who proved outer Pasch implies inner Pasch. Tarski took outer Pasch as an
axiom in [31].

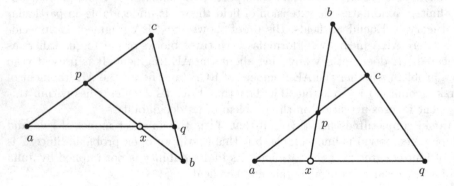

Fig. 3. Inner Pasch (left) and Outer Pasch (right). Line pb meets triangle acq in one
side. The open circles show the points asserted to exist on the other side.

Tarski originally took outer Pasch as an axiom, but following the "final"
version of Tarski's theory in [29], we take inner Pasch. Seeking a quantifier-free
formulation, we introduce a function symbol ip to produce the intersection points
from the five points labeled in the diagram. When the betweenness relations in
the diagram are not satisfied, nothing is asserted about the value of ip on those
points.

4.3 Gupta's Thesis

In his 1965 thesis [12] under Tarski, H. N. Gupta proved two great theorems:[9]

(i) Inner Pasch implies outer Pasch. After that Szmielew's development used inner Pasch as an axiom (A7) and dropped outer Pasch (although Tarski was still arguing decades later [32] for the other choice).

(ii) Connectivity of Betweenness:

$$a \neq b \wedge T(a,b,c) \wedge T(a,b,d) \rightarrow T(a,c,d) \vee T(a,d,c).$$

That is, betweenness determines a linear order of points on a line. Points d and c, both to the right of b on $Line(a,b)$, must be comparable.

The connectivity of betweenness was taken as an axiom by Tarski, but once Gupta proved it dependent, it could be dropped. Gupta never published his thesis, but his proof of connectivity appears as Satz 5.1 in [29]. The proof is complicated: it uses 8 auxiliary points and more than 70 inferences, and uses all the axioms A1-A7. His proof of outer Pasch (from inner Pasch) also occurs in [29] as Satz 9.6.

4.4 Dimension Axioms

(A8) (lower dimension axiom) says there are three non collinear points (none of them is between the other two)

(A9) (upper dimension axiom) says that any three points equidistant from two distinct points must be collinear. In other words, the locus of points equidistant from a and b is a line (not a plane as it would be in $\mathbb{R}^3$).

(A1) through (A9) are the axioms for "Hilbert planes."

4.5 Tarski's Parallel Axiom (A10)

In the diagram (Fig. 4), open circles indicate points asserted to exist. There are other equivalent forms; see [32,29].[10] We would need to introduce new function symbols to work with A10 in a theorem-prover. Since none of the work in this paper depends on the exact formulation of the parallel axiom, we do not discuss alternate formulations.

4.6 Line-Circle Continuity

In Fig. 5, point p is "inside" the circle since $ap \equiv ax$. Then the points indicated by open circles must exist.

---

[9] Gupta got his Ph. D. sixteen years after earning his second master's degree in India. There was at least one more great theorem in his thesis–I do not mean to imply that he proved only two great theorems.

[10] Tarski and Givant [32] label a certain quantifier-free formula "Third Form of Euclid's Axiom", which is misleading, because this formula is not equivalent to A10 (see Exercise 18.4 in [13]).

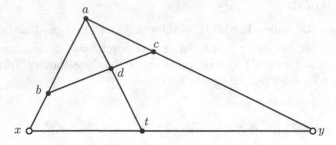

Fig. 4. Tarski's parallel axiom

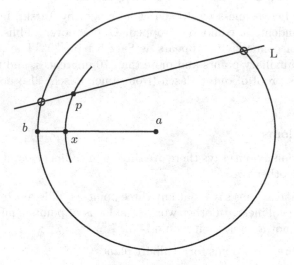

Fig. 5. Line-circle continuity. Line L is given by two points A, B (not shown). The points shown by open circles are asserted to exist.

We use two function symbols ℓc_1 and ℓc_2 to name the points $\ell c_1(A, B, a, x, b, p)$ and $\ell c_2(A, B, a, x, b, p)$, where A and B are two unequal points that determine the line L. In [3], we also introduce another axiom, the point of which is to ensure that the two intersection points occur in the same order on L as A and B. This can be expressed as a disjunction of several betweenness statements, essentially listing the possible allowed orders of the four points. Since non-strict betweenness is used in this axiom, the points p and x might be on the circle, in which case the line is tangent to the circle and the two intersection points coincide. The additional axiom implies that the two intersection points each depend continuously on their arguments. Tarski used an existential quantifier instead of function symbols to formulate line-circle continuity, so the extra axiom was not needed; but we want a quantifier-free axiomatization, and the extra axiom is natural so that there will be one natural model $\mathbb{F}^2$ over each Euclidean field $\mathbb{F}$, instead of uncountably many with strange discontinuous interpretations of ℓc_1 and ℓc_2.

4.7 Circle-Circle Continuity

In Fig. 6, points p and q on circle K are "inside" and "outside" circle C, respectively, because $ax \equiv ap$ and $ax \equiv aq$. Then points 1 and 2 (indicated by open circles in the figure) exist, i.e. lie on both circles. The two intersection points can coincide in some degenerate cases, but if the two circles coincide, so there are more than two intersection points, then points 1 and 2 become "undefined", or technically, since we are not using the logic of partial terms, they just become unknown points about which we say nothing.

As for line-circle continuity, we introduce two Skolem functions cc_1 and cc_2 to define the intersection points. Since there are only point variables, circle C in the figure will be given by its center a and the point y, and circle K is given by its center b and the point q. Hence the arguments of the two Skolem functions are just the points labeled with letters in Fig. 6.

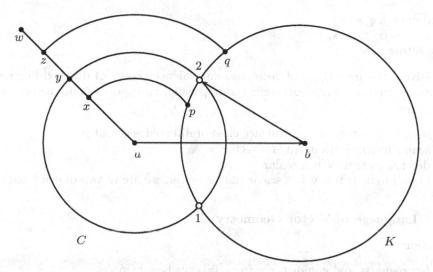

Fig. 6. Circle-circle continuity. p is inside C and q is outside C, as witnessed by x, y, and z, so the intersection points 1 and 2 exist.

In general, when one introduces a Skolem function, one may lose completeness (perhaps that is why Tarski left his axioms in $\forall\exists$ form). Once we Skolemize the circle-circle continuity axiom, we also want extra axioms to distinguish the two points of intersection and ensure that they depend continuously on their arguments. The rule we want to state is that the "turn" from a to b to intersection point 1 is a right-hand turn, and the turn from a to b to intersection point 2 is a left-hand turn. Rather than defining "right turn" directly, we define abc to be a right turn if, when we draw the circles of radius ac and center a, and radius bc and center b, their first intersection point is c; if c is instead the second intersection point, then abc is a left turn (by definition). Then we add an axiom saying that if c and d are on the same side of the line through a and b, and abc is a right turn, so is abd, and the same for left turns. For the definition of "same side", we follow [29], Definition 9.7, p. 71. Since these axioms play no role in this paper, we refer the interested reader to [3] for further details.

5   VG, A Formal Theory of Vector Geometry

In this section, we describe a first-order theory **VG** that contains both geometry and algebra. This theory permits us to formalize the relationships between algebra and geometry in both directions, and to formalize the Chou area method directly. The theories of algebra and of geometry become fragments of **VG**.

5.1   Language of Vector Geometry

Three sorts:

- points p, q, a, b
- scalars α, β, λ, s, t
- vectors **u**, **v**

Intuitively you may think of vectors as equivalence classes of directed line segments under the equivalence relation of parallel transport. Constructors and accessors:

- $p \circ q$ is a vector, the equivalence class of directed segment pq.
- scalar multiplication: $\lambda\mathbf{u}$ is a vector
- dot product: $\mathbf{u} \cdot \mathbf{v}$ is a scalar
- cross product: $\mathbf{u} \times \mathbf{v}$ is a scalar (not a vector, we are in two dimensions)

5.2   Language of Vector Geometry

Relations:

- betweenness and equidistance from Tarski's language
- Equality for points, equality for vectors, equality for scalars. Technically these are different symbols.
- $x < y$ for scalars.

Function symbols (other than constructors and accessors) and constants:

- Skolem symbols for Tarski's language, specifically ext for segment extension, ip for inner Pasch, and Skolem symbols for line-circle and circle-circle continuity, as described above.
- $0, 1, *, +, /$, unary and binary $-$, and $\sqrt{\ }$ for scalars.
- **0** is a vector; $\mathbf{u} + \mathbf{v}$, $\mathbf{u} - \mathbf{v}$, and $-\mathbf{u}$ are vectors.
- $\hat{0}$, $\hat{1}$, and $\hat{i}$ are unequal points.
- $\hat{i}$ is a point equidistant from 1 and from $-1 = ext(1, 0, 0, 1)$.

5.3 Division by Zero and Square Roots of Negative Numbers

$1/0$ is "some scalar" rather than "undefined", because we want to use theorem provers with this language and they do not use the logic of partial terms. One cannot prove anything about $1/0$ so it does not matter that it has some undetermined value. For example, we have the axiom $x \neq 0 \to x * (1/x) = 1$, not the axiom $x * (1/x) = 1$. Other "undefined" terms are treated the same way.

5.4 Axioms of Vector Geometry VG

- Tarski's axioms for ruler-and-compass geometry.
- The scalars form a Euclidean field.
- The obvious axioms for unary $-$ and $/$ and $\sqrt{\ }$ permit us to avoid existential quantifiers in the axioms for Euclidean fields. The use of binary $-$ is a convenience; the axiom is $a - b = a + (-b)$.
- The vectors form a vector space over the scalars.
- The usual laws for dot product and 2d cross product.
- $a \circ b = -b \circ a$
- $p \circ p = \mathbf{0}$
- $E(\hat{0}, \hat{i}, \hat{0}, \hat{1})$
- $E(\hat{i}, \hat{1}, \hat{i}, -\hat{1})$ where $-\hat{1} = ext(\hat{1}, \hat{0}, \hat{1}, \hat{0})$
- If ab and cd are parallel and congruent, then $a \circ b = \pm c \circ d$, with the sign positive if segment ad meets segment bc, and negative if it does not.
- If a, b, c, and d are collinear and ab and cd are congruent, then $a \circ b = \pm c \circ d$, with the appropriate sign (given by betweenness conditions expressing the intuitive idea that the sign is positive if the directed segments ab and cd have the same direction).

The geometrical part (based on Tarski's axioms A1-A10 and line-circle and circle-circle continuity, but in a quantifier-free form) we call "Euclidean geometry" **EG**. There are function symbols for the intersection points of two lines, of a line and a circle, and of two circles; that is a slightly different choice of function symbols from the quantifier-free version of Tarski's axioms used in this paper. See [3] for a detailed formulation.

5.5 Analytic Geometry in VG

This corresponds to the translations across the top and bottom of our (supposedly) commutative diagram. Let ϕ be a formula of **EG**. Let ϕ^* be a translation of ϕ into Euclidean field theory (expressed using scalar variables in **VG**). In fact, more generally we can define ϕ^* when ϕ is a formula of **VG**, not just of **EG**.

The first thing to notice here is that there is more than one way to define such a translation ϕ. The obvious one is the one that is taught to middle-school children, which we call the "Cartesian translation." Lines are given by linear equations, and points by pairs of numbers. In practice this leads to many case distinctions as vertical lines require special treatment. Another translation, invented by Chou

[8], takes a detour through vectors, but can be expressed directly in **VG**, and then (whichever way we interpreted points), of course vectors can be expressed by coordinates using the scalars of **VG**. We will discuss the Chou translation in more detail below. The important fact, expressing the adequacy of **VG** for this part of the method, is the following theorem.

Theorem 1 (Analytic Geometry). *Let ϕ be a formula of* **VG**. *Let $\phi^\star$ be either the Cartesian translation of ϕ, or the Chou translation. Then*

$$\mathbf{VG} \vdash \phi \qquad implies \qquad \mathbf{EF} \vdash \phi^\star$$

where **EF** *is the theory of Euclidean fields.*

Remark. This corresponds to the top of the commutative diagram.

Proof. Here we discuss only the Cartesian translation. By "the x-axis" we mean the line containing α and β, where α and β are the two distinct points mentioned in the dimension axioms. Next we define the Cartesian interpretation $\phi^\star$ of ϕ. To define $\phi^\star$, we have to first assign a term $t^\star$, or a pair of terms (t_1, t_2), of **EF** for each term t of **EG**. In fact, we define $t^\star$ for all terms of **VG**, not just **EG**. Since points and vectors are to be interpreted as pairs of scalars, we use pairs for terms of those types; for a term of **EF** we just have $t^\star = t$. Otherwise we write $t^\star = (t_1^\star, t_2^\star)$. When t is a point variable x, then x occurs in the official list of all point variables as the n-th entry, for some n, and we define $x^\star$ to be the pair consisting of the scalar variables occurring with indices $4n$ and $4n+1$ in the official list of scalar variables. Similarly for vector variables, but using indices $4n + 2$ and $4n + 3$.

The definitions of $\phi^\star$ for the case of atomic formulas involving betweenness or segment congruence is straightforward analytic geometry. For details see [13] or [3]. To extend the definition of $\phi^\star$ from **EG** to **VG**, we have to show that dot product and cross product of vectors can be algebraically defined. For example,

$$(t \times s)^\star = (t_1^\star s_2^\star - t_2^\star s_1^\star)$$

Once $t^\star$ is defined, then $\phi^\star$ commutes with the logical connectives and quantifiers, except that for quantifiers, each point or vector variable is "doubled", i.e. changed to two scalar variables.

Now the theorem is proved by induction on the length of proofs in **VG**. The base case is when ϕ is an axiom of **VG**. We verify in **EF** that the algebraically defined relations of betweenness and equidistance satisfy the axioms of **EG**. That is lengthy, and not entirely straightforward, in the case of circle-circle continuity (but note that the same complications occur whether we are using model theory or interpretations). See, for example, [13], page 144. See also [3] for some details omitted in [13]. We note that if ϕ is quantifier-free (or AE) in EG, then $\phi^\star$ is also quantifier-free (or AE).

5.6   Geometric Arithmetic in VG

Along the bottom of the commutative diagram, we have a translation ϕ° from **EF** to **EG**, following Descartes with improvements by Hilbert. We show that this translation can be extended to be defined on terms and formulas of **VG**, not only of **EF**. We fix a particular line L (given by the two unequal points α and β, which we use as the interpretations of 0 and 1, respectively). Scalars of **VG** are interpreted as points on L, i.e. collinear with α and β. Vectors are interpreted as pairs of such points. Since there is no pairing function in **EG**, the formula ϕ° may have more variables than ϕ, as each vector variable converts to two point variables. The terms $(t+s)^\circ$, $(t-s)^\circ$, $(-t)^\circ$, $(t \cdot s)^\circ$, and $(\sqrt{t})^\circ$, are defined using terms of **EG**. For example, the definition of $(t \cdot s)^\circ$ should be the one suggested by Fig. 2.

It is not at all obvious that t° can be defined using the terms of **EG**, which do not include a symbol for definition-by-cases; in other words there are no terms in **EF** to construct a term $d(a, b)$ that is equal to α if $a = b$ and to β otherwise. But the definitions of addition and multiplication given by Descartes require such a case distinction; Hilbert's multiplication does not, but his addition still does. We have shown in [3] that an improved (continuous) definition of addition can be given; there only constructive logic is used, but here that is not an issue. If that were not true, we would simply include a definition-by-cases symbol in **EG**, and our main results would not be affected; but it is not necessary. For complete details of the interpretation from **EF** to **EG**, see [3].

Theorem 2 (Geometric algebra). *Let ϕ be a formula of **EG**. Let ϕ° be the translation discussed above of **VG** into **EG**. Then*

$$\mathbf{VG} \vdash \phi \qquad implies \qquad \mathbf{EG} \vdash \phi^\circ$$
$$\mathbf{VG} \vdash \phi \qquad implies \qquad \mathbf{EG} \vdash (\phi^\star)^\circ$$

Remark. This corresponds to the bottom of the commutative diagram.

Proof Sketch. It has to be verified geometrically that multiplication, addition, and square root satisfy the laws of Euclidean field theory. This goes back to Descartes and Hilbert, but as noted above, since we do not have a test-for-equality function in **EG**, a more careful definition of addition is required. Since we have extended the interpretation to **VG**, we also need to verify the laws of vector spaces and of cross product and dot product geometrically. It is possible to do this directly, but we can also circumvent the need for those details, by defining ϕ° to be $(\phi^\star)^\circ$ for formulas ϕ involving vectors. Note that the language of **VG** has no terms constructing vectors from points, so if ϕ contains terms of type vector, it does not contain betweenness or equidistance or any subterms of type point. Hence it is not actually necessary to directly verify the laws of cross product and dot product and vector spaces geometrically.

5.7   Conservativity and Commutativity

Suppose we start with a geometric theorem ϕ and somehow prove either it or $\phi^\star$ with the aid of analytic geometry. (By Theorem 1, if we have a proof of $\phi^\star$, we can get a proof of ϕ, and vice-versa.) Then can we eliminate the "scaffolding" of analytic geometry, and find a purely geometric proof of ϕ? Yes, we can:

Theorem 3. VG *is a conservative extension of* **EG**. *That is, if* ϕ *is a formula in the language of Tarski's geometry* **EG**, *and* **VG** *proves* ϕ, *then* **EG** *proves* ϕ.

Proof. Suppose ϕ is a formula of **EG**, and **VG** proves ϕ. Then by Theorem 2, ϕ° is provable in **EG**. But ϕ° is ϕ since ϕ is a formula of **EG**. That completes the proof of the theorem.

Suppose we start with a formula ϕ of Euclidean field theory, and find a proof of it using vectors, or even using the full apparatus of **VG**. Then it is already provable from the axioms of Euclidean fields:

Theorem 4. VG *is a conservative extension of Euclidean field theory* **EF**.

Proof. Suppose ϕ is a formula of Euclidean field theory **EF**, and suppose **VG** proves ϕ. Then $\phi^\star$ is provable in **EF**. But by definition of $\phi^\star$, when ϕ is a formula of **EF**, $\phi^\star$ is exactly ϕ. Hence **EF** proves ϕ. That completes the proof.

Next we show that the two interpretations $\phi^\star$ and ψ° are, up to provable equivalences, inverses. This is by no means immediate, since the definitions have no apparent relation to one another. But nevertheless, they both express the same geometric relationships.

Theorem 5 (Commutativity). *Let* ϕ *be a formula of* **EG** *with no free variables. Then*

$$(\phi^\star)^\circ \leftrightarrow \phi$$

is provable in **EG**. *Similarly, if* ψ *is a theorem of* **EF** *with no free variables, then*

$$(\psi^\circ)^\star \leftrightarrow \psi$$

is provable in **EF**.

We will not give a proof of this theorem here, as it is highly technical. The theorem has to be first formulated in a way that holds for formulas with free variables, as well as for formulas without, and then proved by induction on the complexity of ϕ. The technical issue here is the "doubling" of variables when we pass from point variables to scalar variables, and the "uncoordinatizing" in the other direction. We will explain what "uncoordinatizing" means next.

We will illustrate the proof by explaining one example, the case when ϕ is $T(\alpha, y, \beta)$. Then the point variable y is "doubled" by $\phi^\star$ to the two scalar variables (λ_1, λ_2), intuitively representing the coordinates of y. (Defining this precisely depends on setting up a one-to-two correspondence between the lists of variables of type point and type scalar.) Then $\phi^\star$ is

$$\lambda_2 = 0 \wedge 0 \leq \lambda_1 \wedge \lambda_1 \leq 1.$$

Now going back to geometry, the two scalar variables do not convert back to one point variable. Instead they become two variables of type point, y_1 and y_2, restricted (by $(\phi^\star)^\circ$) to lie on the x-axis (the line through α and β). We write $Col(x, y, z)$ for "x, y, and z lie on the line containing distinct points x and y", defined in terms of betweenness. Then $(\phi^\star)^\circ$ is equivalent to (though not literally the same as)

$$Col(\alpha, \beta, y_1) \wedge Col(\alpha, \beta, y_2) \wedge y_2 = \alpha \wedge T(\alpha, y_1, \beta).$$

Finally we construct the point $q = F(y_1, y_2)$ using the F described above. Then $X(q) = y_1$ and $Y(q) = y_2$. Then

$$(\phi(y)^\star)^\circ \leftrightarrow ((\phi^\star)(\lambda_1, \lambda_2))^\circ$$
$$\leftrightarrow \phi(q)$$
$$(\phi(y)^\star)^\circ \leftrightarrow \phi(F(y_1, y_2)) \tag{1}$$

Here y_1 and y_2 are point variables related to the original point variable y by this rule: if y is the n-th point variable x_n, then y_1 is x_{2n} and y_2 is x_{2n+1}. Now the variables on the left side of (1) are not related to the variables on the right in any semantic way; to state the commutativity theorem for ϕ we need this:

$$y = F(y_1, y_2) \to ((\phi(y)^\star)^\circ \leftrightarrow \phi(y)) \tag{2}$$

where in spite of appearances the formula $((\phi(y)^\star)^\circ$ contains y_1 and y_2 free, not y. Equation (2) demonstrates the way that $\phi \leftrightarrow (\phi^\star)^\circ$ is generalized to formulas with free variables. Granted, this is technical, but it works and is in some sense natural, and it is the price we have to pay for the benefits of an explicit interpretation. Once this is correctly formulated, the rest of the proof is straightforward (which is not the same thing as "short").

5.8   Algebra in VG

When we "prove" a geometric theorem by making a corresponding algebraic computation, we have not yet proved anything; we have only made a computation. In order to convert such a computation (ultimately) to a geometric proof, it will first be necessary to convert an algebraic computation to an algebraic proof. This also goes under the name of "verifying a computation", or sometimes, "verifying the correctness of a computation."

What we want to prove is that "computationally equal" terms t and s are provably equal. For example, we can compute by simple algebra that

$$(x^2 - y^2)^2 + 4x^2y^2 = (x^2 + y^2)^2$$

But that does not deliver into our hands a formal proof of that equation from the axioms of Euclidean field theory.

The principal problem here is that "computationally equal" is not very well defined. One is at first tempted to say: if your favorite computer algebra system says the terms are equal, they are computationally equivalent. But if we take *that* definition, then it is false that computationally equivalent terms are provably equal. For example, Sage and *Mathematica* agree that $x \cdot (1/x) = 1$, but that is not provable in **EF**, since, if it were, we would have $0 \cdot 1/0 = 1$, but since $0 \cdot z = 0$ we also have $0 \cdot 1/0 = 0$, hence $1 = 0$, but $1 \neq 0$ is an axiom of **EF**.

Such problems arise from the axioms of **EF** that are not equational, for example $x \neq 0 \rightarrow x \cdot (1/x) = 1$ and $x \geq 0 \rightarrow (\sqrt{x})^2 = x$. If we use these equations without regard to the preconditions, false results can be obtained.

We do not know how to define "computationally equivalent terms" except by provability of $t = s$ in **EF**, or more generally, the vector and scalar part of **VG**. The search for a theorem or general result degenerates to a practical problem: given a computation by a computer algebra system that $t = s$, determine the minimal "side conditions" ϕ on the variables of $t = s$ necessary for the provability of $t = s$ and find a first-order proof of $t = s$. One may or may not wish to consider paramodulation steps as legal.

5.9 Chou's Method Formalizable in VG

This subsection presumes familiarity with Chou's method [8]. Readers without that prerequisite may skip this subsection and continue reading, but since Chou's method is an important method of proving geometrical theorems by algebraic computations, we want to verify that it can be directly formalized in **VG**.

We start with Chou's basic concept, the *position ratio*. In **VG**, we define

$$\frac{ab}{cd} := pr(a, b, c, d) = \frac{(a \circ b) \cdot (c \circ d)}{(c \circ d) \cdot (c \circ d)}$$

Our pr is defined whenever $c \neq d$. Chou's position ratio is defined only when a, b, c, and d are collinear, but in that case they agree.

Chou makes extensive use of the signed area of an oriented triangle. We define that concept in **VG** by

$$\mathcal{A}(p, q, r) := \frac{1}{2}(q \circ p) \times (q \circ r).$$

Chou's other important concepts and theorems can also be defined and proved in **VG**. In particular, the co-side theorem can be proved in **VG**. This should be checked by machine; it would make a good master's thesis.

6   Finding Formal Proofs from Tarski's Axioms

An essential part of the project of making our diagram commute is to find formal proofs in geometry up to the point where multiplication, addition, and square root can be geometrically defined and their field-theoretic properties proved. We chose to attack this project using Tarski's axioms and resolution theorem provers. One of the reasons for this choice is the existence of a very detailed "semiformal" development due to Szmielew.[11] Another reason is that others have tried in the

---

[11] Wanda Szmielew developed course notes for her course in the Foundations of Geometry at UC Berkeley, 1965-66, and gave a copy to her successor as instructor of that course, Wolfram Schwäbhauser. These notes incorporated important contributions from Gupta's thesis [12], including the two theorems mentioned above. After her death, her notes were published with "inessential changes" as part of [29].

past to work with resolution theorem provers and Tarski's axioms. Specifically, MacPharen, Overbeek, and Wos [18] worked 37 years ago with Tarski's 1959 system; after the publication of [29], Quaife [25,26] used Otter to formalize the first four of the fifteen chapters of Szmielew's development (that is, Chapters 2-5 out of 2-16). Quaife solved some, but not all, of the challenge problems from [18], and added some challenge problems of his own. In 2006, Narboux [20] checked Szmielew's proofs using Coq, up through Chapter 12. While this is fine work, the fact remains that after almost forty years, we have collectively still not produced computer-checked proofs of Szmielew's development–whether by an automated reasoning program (such as Otter), an interactive proof-checker (such as Coq), or by any other means.[12]

Wos and I undertook to make another attempt. One aim of this project is to produce a formal proof of each theorem ("Satz") in Szmielew, using as hypotheses (some of) the previous theorems and definitions, with the ultimate aim of formalizing the definitions and properties of multiplication, addition, and square root.

A second aim of the project is to see how much the techniques available for automated deduction have improved in the 20 years since Quaife. Would we now be able to solve the challenge problems that were left unsolved at that time?

And ultimately, a third aim of the project is to reach the propositions of Euclid, but based on Tarski's axioms.

6.1 Szmielew in Otter

Larry Wos and I experimented with going through Szmielew's development, making each theorem into an Otter file, giving Otter the previously proved theorems to use. We also used Prover9 sometimes, but we found it did not perform noticeably better (or worse) on these problems. Our aim was to obtain Otter proofs of each of Szmielew's theorems.

Some basic facts about Otter (and Prover9) will be helpful. Otter gives a weight to every formula (by default, the total number of symbols). You can artificially adjust the weights using a list called the "weight list". There is a parameter called **max_weight**; formulas with larger weights are discarded to keep the search space down. Thus you can control the search to some degree by assigning certain formulas low weights and finding a good value of **max_weight** that allows a proof to be found: just large enough that all needed formulas are kept, not so large that the prover drowns in irrelevant conclusions. These remarks apply to both Otter and Prover9; the difference between the two provers lies in the algorithm for choosing the next clauses to be considered for generating new clauses.

One technique we used is called "giving Otter the diagram." This means defining a name for each of the points that need to be constructed. For example,

---

[12] William Richter has also checked Szmielew up through Satz 3.1 in miz3 (see http://www.math.northwestern.edu/~richter/TarskiAxiomGeometry.ml).
He has also checked some proofs from Hilbert's axioms; the code is in hol_light/RichterHilbertAxiomGeometry/ in the HOL-Light distribution.

if the diagram involves extending segment ab beyond b by an amount cd, you would add the line $q = ext(a, b, c, d)$. The point of doing so is that q, being an atom, gets weight 1, so terms involving q will be more likely to be used, and less likely to be discarded.

We also used "hints" and "resonators" [34]. By this we mean the following:

(i) We put some of the proof steps (from the book proof) in as preliminary goals. We get proofs of some of them.

(ii) We put the steps of those proofs into Otter's weight list, giving them a low weight, to ensure that they be chosen quickly to make new deductions. The bound `max_weight` is then set to the smallest value that will ensure their retention. This prevents the program from drowning in new and possibly irrelevant conclusions.

The technique of resonators can be used in other ways as well. For example, if you are trying to prove C and you have proofs of some lemma A and a proof of C from the assumption A, but you cannot directly get a proof of C, then put the steps of the two proofs you do have in as resonators. Very likely you will find the desired proof of C. Wos has also used resonators very successfully to find shorter proofs, once a long proof is in hand.

6.2   What Happened

Our first observation was that it is necessary to give Otter the diagram, in the sense described above. Once we started doing that, we went through Chapters 2 and 3 (of 2–16) rapidly and without difficulty.

We hit our first snag at Satz 4.2. An argument by cases according as $a = c$ or $a \neq c$ is used. Otter could do each case, but not the whole theorem! We tried Prover9. Prover 9 could prove Satz 4.2, but it took 67,000 seconds! (1 day = 86,400 sec.)

The inability to argue by cases is a well-known problem in resolution theorem-proving. On perhaps ten (out of more than 100) subsequent theorems, we had to help Otter with arguments by cases. Sometimes we did that by putting in the case split explicitly, and giving the cases low weights. For example we would put in `b=c | b!=c` and then give both literals a negative weight. With this trick, if the cases can be done in separate runs, we could sometimes get a proof in a single run.[13] If not, then we used the proof steps of both cases as resonators.

Chapter 5 of [29] contains a difficult theorem from Gupta's thesis [12], the connectivity of betweenness (Satz 5.1). That theorem is

$$a \neq b \wedge T(a, b, c) \wedge T(a, b, d) \rightarrow T(a, c, d) \vee T(a, d, c).$$

---

[13] Ross Overbeek suggested a general strategy: if you don't get a proof, look for the first unit ground clause deduced, and argue by cases (in two runs) on that clause. That strategy would have worked on Satz 4.2. Here is a project: implement this strategy using parallel programming.

Neither Otter nor Prover9 could prove Gupta's theorem without help. We used resonators, starting with about thirty of Gupta's proof steps. This technique was successful. We found a proof of Satz 5.1.

After proving the connectivity of betweenness, we had no serious difficulties with the rest of Chapters 5 and 6; Otter required no help except a couple of case splits.

In 1990, Quaife made a pioneering effort (using Otter) to find proofs in Tarski's geometry. He used the version of Tarski's axioms [29], just as we do. Quaife made it a bit farther than where Wos and I hit our first snag in Szmielew. Most of Quaife's theorems are in Szmielew Chapters 2 and 3, or the first part of 4, or are similar to such theorems, but use some defined notions such as "reflection," which occurs in Chapter 7 of [29]. His most difficult example was that the diagonals of a "rectangle" bisect each other. Here a "rectangle" is a quadrilateral with two opposite sides equal and the diagonals equal. This theorem is weaker than Lemma 7.21 in [29], which says that in a quadrilateral in which (both pairs of) opposite sides are congruent, the diagonals bisect each other.

Quaife left four challenge problems, which are theorems in that part of [29] that we formalized. Although he did not say so explicitly, it is clear that these should be solved from axioms A1-A9, i.e. without the parallel axiom or any continuity assumptions. We list them here:

- the connectivity of betweenness (Satz 5.1 in [29])
- every segment has a midpoint (Satz 8.22 in [29]).
- inner Pasch implies outer Pasch (Satz 9.6 in [29])
- Construct an isosceles triangle with a given base (an immediate corollary of Satz 8.21 and Satz 8.22, the existence of midpoints and perpendiculars).

The difficulty of proving the existence of a midpoint lies in the fact that no continuity axioms are allowed. (The usual construction using two circles is thus not applicable.) This requires developing the theory of right angles and perpendiculars, and takes most of Chapters 7 and 8 of [29]. The construction depends on two difficult theorems: the construction of a perpendicular to a line from a point not on the line, and the construction of a perpendicular to a line through a point on the line (Satz 8.18 and 8.20). These in turn depend on another theorem of Gupta, called the "Krippenlemma" (Lemma 7.22). The proof that inner Pasch implies outer Pasch is one of the highlights of [12].

6.3 Our Results

We were eventually able to prove all four of Quaife's challenge problems, and indeed all the results from [29] up to and including Satz 9.6. The proofs we found, the Otter input files to produce them, and some discussion of our techniques, are available at [2]. Satz 7.22 (the Krippenlemma) and Satz 8.18 (construction of the perpendicular) were extremely difficult, and required many iterations of proofs of intermediate results, and incorporation of new resonators from those proofs. We could never have found these Otter proofs without the aid of Gupta's proofs, so in some sense this is "computer-assisted deduction", intermediate between "proof-checking" and "automated deduction."

Why were we able to do better in 2012 than Quaife could do in 1990? Was it that we used faster computers with larger memories? No, it was that we used techniques unknown to Quaife. We could not find these proofs with 2012 computers using only Quaife's techniques. Maybe we could have found the proofs we found with 1990 computers and 2012 techniques, but we're glad we didn't have to. Quaife knew how to tell Otter what point to construct (and we learned the technique from him), but he didn't know about resonators.

6.4   Euclid from Tarski

As of summer 2012, neither by hand nor by machine had development from Tarski's axioms reached the first proposition of Euclid, more than half a century after Tarski formulated his axioms, although an approach from a far less parsimonious axiom set has allowed the mechanization of some of Euclid [1]. Quaife did not get as far as proving any theorem about circles. Neither did Szmielew or Gupta. All these authors wanted to postpone the use of even line-circle or circle-circle continuity as long as possible, while Euclid uses it (implicitly) from the outset. We felt that it was high time to prove at least the first proposition of Euclid from Tarski's axioms.

Euclid's Book I, Prop. 1. constructs an equilateral triangle, as shown in Fig. 7. The open circle indicates the constructed point. Euclid's proof does not meet the modern standards of rigor, according to which one would need some sort of continuity axiom and perhaps some sort of dimension axiom to prove Prop. 1, since if the circles were in different planes, they would not meet. It turns out that the dimension axioms are not needed, because "circle-circle continuity" is sphere-sphere continuity in $\mathbb{R}^3$.

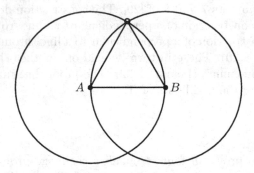

Fig. 7. Euclid Book I, Proposition 1

The result: Otter proves Euclid I.1 from circle-circle continuity in less than two seconds, with a good choice of inference rules. When we first did it, it took eleven minutes, which is about how long it will take you by hand.

Euclid's Prop I.2 says that given three points A, B, and C, a point D can be constructed such that $AD = BC$. That is immediate from Tarski's segment extension axiom (A4). Euclid only postulated you can extend a segment *somehow*, so in a sense, Tarski's (A4) is unnecessarily strong.

Euclid's Prop I.3 mentions the concepts "the greater" and "the lesser" between segments. In Tarski's primitives, we would define $ab \leq cd$ to mean that for some x we have $ab \equiv cx$ and $T(c, x, d)$. Then Prop. I.3 has no content; in other words Prop. I.3 amounts to a definition of "the greater" and "the lesser", which in Euclid are "common notions."

Prop. I.4 is the SAS congruence criterion. That requires defining angle congruence; it is Satz 11.4 in Szmielew! There is a big jump from the first three propositions to Prop. I.4. The reason is that angle congruence and indeed comparison of angles ($\leq$ for angles) are primitive in Euclid, but defined in Tarski's system. The resulting complications have little to do with automated deduction. They are the consequence of choosing a very parsimonious formal language. Therefore, we should complete Szmielew Chapter 11 first, or take some axioms about comparison and congruence of angles, in order to formalize Euclid directly. (That is the approach taken in [1].) We plan to return to Euclid after finishing the formalization of Szmielew up to Chapter 11. For example, Euclid's Prop. I.8 is the SSS angle criterion, Satz 11.51 in Szmielew.

7   Proof by Computation, in Theory and Practice

In this section, we discuss the top and right sides of the diagram. The plan, in theory, is to verify the truth of (which in a loose sense is to prove), some geometric formula A. We start by expressing it as a system of algebraic equations (or inequalities) using analytic geometry (or by some other method), introducing new variables for the coordinates of the points to be constructed (or some other quantities depending on those points). Then we calculate to see if these equations can be satisfied. If the calculation succeeds, then A is verified. But we still do not have a first-order proof of A.

To put this method into practice, we need to answer two questions: Exactly how will we convert from geometry to algebra, and exactly how will we make the required computations? Among the ways to convert geometry to algebra, we mention the ordinary introduction of coordinates, and Wu's method [35], and Chou's area method [8]. Among the ways to compute, we mention Gröbner bases and the Collins CAD algorithm [7,6]. While theoretically, any geometry problem can be solved by CAD, since it is a decision procedure for real-closed fields, in practice, it breaks down on problems with five or six (number) variables, so a geometry problem with four points is likely to be intractable, and geometry problems with fewer than four points are rare. On the other hand, Wu's method and Chou's area method have been used to prove hundreds of beautiful theorems, some of them completely new. In that sense, they far outperformed resolution theorem proving.

In spite of the dramatic successes of these methods, we point out two short-comings. First, both these methods work only on theorems that translate to algebra using equations, with no inequalities. Thus the "simple" betweenness theorems of Szmielew Chapter 3 are out-of-scope.

Second, you cannot ask for a proof from ruler-and-compass axioms (or indeed from any geometric axioms at all). You can only ask if the theorem is true in $\mathbb{R}^2$. Thus there is no problem trisecting an angle; this is not about ruler-and-compass geometry. A proposition like Euclid I.1 is just trivial: all the subtleties and beauties of the first-order proof are not captured by these methods. It just computes algebraically that there is a point on both circles.

In short, when using these methods, we are not doing geometry. We are doing algebra. It is these shortcomings that we propose might be rectified, if we could make the diagram commute in practice. Then we could go along the bottom of the diagram, benefiting from computation on the right, and still end up with geometrical proofs on the left.

In theory, we should be able to get geometric proofs by going across the bottom of the diagram from right to left. That is, to convert the algebra performed by Chou's method into first-order proofs of algebraic theorems, from some algebraic axioms, and then back-translate to geometry, using the geometric definitions of addition and multiplication. In theory this can certainly be done. In practice, the authors of [8] were aware of this possibility, and discuss it on pp. 59–60, but they say, "The geometric proofs produced in this way are expected to be very long and cumbersome, and as far as we know no single theorem has been proved in this way." Nevertheless, those proofs, if we could find them, would be proofs and not just computations.

8   From Computation to Proof: Going around the Dragons

Here is the plan to find a first-order proof of a given geometric theorem by going across the top of the diagram, down the right, and back, all within **VG**.

- Start with a geometric theorem ϕ to be proved.
- Do the analytic geometry to compute $\phi^\star$. (By Chou or Descartes)
- Find (e.g. by Chou's program or by hand) an informal proof that $\phi^\star$ is true, by calculation.
- Get a formal proof in **VG** of $\phi^\star$, i.e., verify the calculation.
- Use (an implementation of) Theorem 1 to get a proof of ϕ in **VG**.
- Eliminate the non-geometrical axioms to get a proof of ϕ. This can be done, at least in theory, by (an implementation of) Theorem 3.

The main point to be made about this plan is that the difficulty is essentially a "boot-strapping" issue. To get started, we need (machine) formalization of the geometric definitions of addition, multiplication, and square root. These proofs need to be produced just once, and then we can use them to find proofs of many different geometrical theorems. The central importance of the theorems justifying these definitions has long been recognized, as these theorems are in some sense the culminating results of both [14] and [29]. Indeed, the key to proving the properties of multiplication (no matter whether one uses the definition of Descartes or that of Hilbert) is the theorem of Pappus (or Pascal as Hilbert called it). Even the commutativity of addition is not completely trivial. Chapter 15 of Szmielew [29] has the details.

Narboux [20] tried formalizing [29] in Coq, but he didn't get to Chapter 15. Wos and I tried it with Otter, as reported here, but we didn't get to Chapter 15 (yet) either. It turned out that using Otter was not as efficient in finding formal proofs as we had hoped, many human hours were also required. Our hope that every theorem in [29] would be a single, easy run with Otter turned out not to be justified; while that was true of the simpler theorems, every theorem complex enough to require a diagram required several runs, case distinctions made by hand, points defined, and the use of resonators made from lemmas or partial results. As mentioned above, Coq does not produce first-order proofs, and it is probably not easy to extract them from Coq proofs.

9   A Test Case: The Centroid Theorem (Medians All Meet)

We propose a test case for the back-translation method, once someone manages to formalize the definitions of addition, multiplication, and square root. Namely, the theorem that all the medians of a triangle meet in a single point; this is known as the "centroid theorem." Perhaps it is possible to prove this theorem formally from Tarski's axioms using theorems of [29], but that would not count as a solution of this test case.

When Chou's area method is applied to this example, the computations are quite simple (see [8], p. 12); even Cartesian analytic geometry is not very complicated. What is required are the following steps:

 – Formalize the geometry-to-algebra reasoning in VG.
 – Formalize the algebraic computation in VG.
 – Carry out the back-translation and get a formal proof in EG.

That proof would no doubt be long and not very perspicuous. One of the reasons we chose Otter, is that at this point, we could shorten that long proof, using Wos's proof-shortening techniques. Perhaps the complications would melt away, leaving a short proof. Or perhaps the proof would remain impenetrable; we cannot know without performing the experiment. We note, however, that even if the long proof is obtained by another tool, it could still be translated into Otter's language for attempts at proof-shortening.

10   Four Challenge Problems

Having solved some of Quaife's challenge problems, we offer four more. Of course, it goes (almost) without saying that it is a challenge to finish the formalization of [29]; we mean four *in addition* to that. Our first three challenge problems involve line-circle continuity (**LC**) and circle-circle continuity (**CC**). These three problems are

$$\mathbf{CC} \to \mathbf{LC} \qquad \text{using A1-A9 only}$$
$$\mathbf{LC} \to \mathbf{CC} \qquad \text{using A1-A10}$$
$$\mathbf{LC} \to \mathbf{CC} \qquad \text{using A1-A9 only}$$

A first-order proof that **CC** implies **LC** using A1-A9 is sketched on pages 200–202 of [11]. The other direction, **LC** $\to$ **CC**, is more difficult. If we allow the use of the parallel axiom A10, then it is relatively easy to prove that implication model-theoretically. What has to be shown is that if $\mathbb{F}$ is a Pythagorean field, and $\mathbb{F}^2$ satisfies either one of the line-circle or circle-circle continuity, then $\mathbb{F}$ is a Euclidean field. This is done by ordinary analytic geometry; see [13], p. 144, with missing details supplied as in [3]. But that still doesn't give us a first-order proof. With the aid of the parallel postulate (A10), the proof by analytic geometry could, in theory, be used with back-translation to get a first-order proof. In practice, we do not expect to be able to convert this model-theoretic proof to first-order in the near future, so it is a challenge to find a first-order proof by some other means.

In [30], Strommer showed that **LC** $\to$ **CC** can be proved without the parallel axiom. A model-theoretic proof, based on the Pejas classification of Hilbert planes [23], is also known; see the discussions in [11], p. 202 and [13], p. 110. Strommer's proof has the advantage of being first-order (although it is couched in terms of Hilbert's axiom system, not Tarski's). Strommer's proof can probably be made computer-checkable by proceeding through his paper one theorem at a time, but it is a challenge to do so.

The fourth challenge problem is as follows: *Prove from A1-A10 that it is possible to construct an equilateral triangle on a given base.* That is, prove Euclid I.1 without using circles. Hilbert raised the problem of making such a construction, and gave a solution, in his 1898 "vacation course" [15], page 169. Hilbert's solution is quite simple; it is based only on constructing perpendiculars and bisecting segments. One successively constructs right triangles with hypotenuses of length $\sqrt{2}$, $\sqrt{3}$, $\sqrt{3}/2$, as shown in Fig. 8. Since Chapter 8 of Szmielew has the required constructions of perpendiculars and midpoints, we might be in a position to try to get an Otter proof corresponding to Hilbert's construction. But there is a piece of analytic geometry at the end, involving the Pythagorean theorem, which requires the parallel postulate A10. To get a proof from A1-A10, one will certainly have to use the parallel axiom, because the theorem (as Hilbert knew) is not true in all Hilbert planes, i.e. does not follow from A1-A9. See for example [13], Exercise 39.31, p. 373. We tried this problem by giving Otter the diagram for Hilbert's construction, but so far to no avail. We could only apply a 1990 technique, because we have no idea what resonators to use. If the program suggested in this paper could be carried through, we could back-translate Hilbert's proof from analytic geometry to A1-A10.

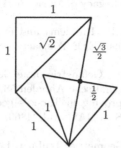

Fig. 8. Constructing an equilateral triangle without using circles

References

1. Avigad, J., Dean, E., Mumma, J.: A formal system for Euclid's Elements. Review of Symbolic Logic 2, 700–768 (2009)
2. Beeson, M.: http://www.michaelbeeson.com/research/FormalTarski/index.php
3. Beeson, M.: Foundations of Constructive Geometry (2012), Available on the author's website, www.michaelbeeson.com/research/papers/pubs.html
4. Beeson, M.: Logic of ruler and compass constructions. In: Cooper, S.B., Dawar, A., Löwe, B. (eds.) CiE 2012. LNCS, vol. 7318, pp. 46–55. Springer, Heidelberg (2012)
5. Borsuk, K., Szmielew, W.: Foundations of Geometry: Euclidean and Bolyai-Lobachevskian Geometry, Projective Geometry. North-Holland, Amsterdam (1960); translated from Polish by Marquit, E.
6. Brown, C.W.: QEPCAD B, a program for computing with semi-algebraic sets using cade. SICSAM Bulletin 37, 97–108 (2003)
7. Caviness, B.F., Johnson, J.R. (eds.): Quantifier Elimination and Cylindrical Algebraic Decomposition. Springer, Wien (1998)
8. Chou, S.C., Gao, X.S., Zhang, J.Z.: Machine Proofs in Geometry: Automated Production of Readable Proofs for Geometry Theorems. World Scientific (1994)
9. Feferman, S. (ed.): The Collected Works of Julia Robinson. American Mathematical Society (1996)
10. Fischer, M.J., Rabin, M.O.: Super–exponential complexity of Presburger arithmetic. SIAM-AMS Proceedings VII, 27–41 (1974); reprinted in [7], pp. 27–41
11. Greenberg, M.J.: Old and new results in the foundations of elementary plane euclidean and non-euclidean geometries. American Mathematical Monthly 117, 198–219 (2010)
12. Gupta, H.N.: Contributions to the Axiomatic Foundations of Geometry. Ph.D. thesis, University of California, Berkeley (1965)
13. Hartshorne, R.: Geometry: Euclid and Beyond. Springer (2000)
14. Hilbert, D.: Foundations of Geometry (Grundlagen der Geometrie). Open Court, La Salle, Illinois (1960), 2nd English edition, translated from the tenth German edition by Unger, L. Original publication date (1899)
15. Hilbert, D.: David Hilbert's lectures on the foundations of geometry 1891-1902. Springer, Heidelberg (2004),edited by Hallett, M and Majer, U.
16. Kempe, A.B.: On the relation between the logical theory of classes and the geometrical theory of points. Proceedings of the London Mathematical Society 21, 147–182 (1890)

17. Marchisotto, E.A., Smith, J.T.: The Legacy of Mario Pieri in Geometry and Arithmetic. Birkhauser, Boston (2007)
18. McCharen, J., Overbeek, R., Wos, L.: Problems and experiments for and with automated theorem-proving programs. IEEE Transactions on Computers C-25(8), 773–782 (1976)
19. Mollerup, J.: Die Beweise der ebenen Geometrie ohne Benutzung der Gleichheit und Ungleichheit der Winkel. Mathematische Annalen 58, 479–496 (1904)
20. Narboux, J.: Mechanical theorem proving in Tarski's geometry. In: Botana, F., Recio, T. (eds.) ADG 2006. LNCS (LNAI), vol. 4869, pp. 139–156. Springer, Heidelberg (2007)
21. Pasch, M.: Vorlesung über Neuere Geometrie. Teubner, Leipzig (1882)
22. Pasch, M., Dehn, M.: Vorlesung über Neuere Geometrie. B. G. Teubner, Leipzig (1926), the 1st edition (1882), which is the one digitized by Scholar, G. does not contain the appendix by Dehn
23. Pejas, W.: Die Modelle des Hilbertschen Axiomensystems der absoluten Geometrie. Mathematische Annalen 143, 212–235 (1961)
24. Pieri, M.: La geometry elementare istituita sulle nozioni di "punto" e "sfera" (elementary geometry based on the notions of point and sphere). Memorie di Matematica e di Fisica della Società Italiana delle Scienze 15, 345–450 (1908), english translation in [7], pp. 160–288
25. Quaife, A.: Automated development of Tarski's geometry. Journal of Automated Reasoning 5, 97–118 (1989)
26. Quaife, A.: Automated Development of Fundamental Mathematical Theories. Springer, Heidelberg (1992)
27. Renegar, J.: Recent progress on the complexity of the decision problem for the reals. DIMACS Series 6, 287–308 (1991); reprinted in [7], pp. 220–241
28. Robinson, J.: Definability and decision problems in arithmetic. Journal of Symbolic Logic 14, 98–114 (1949); reprinted in [9], pp.7–24
29. Schwabhäuser, W., Szmielew, W., Tarski, A.: Metamathematische Methoden in der Geometrie: Teil I: Ein axiomatischer Aufbau der euklidischen Geometrie. Teil II: Metamathematische Betrachtungen (Hochschultext). Springer (1983), reprinted 2012 by Ishi Press, with a new foreword by Beeson, M.
30. Strommer, J.: Über die Kreisaxiome. Periodica Mathematica Hungarica 4, 3–16 (1973)
31. Tarski, A.: What is elementary geometry? In: Henkin, L., Suppes, P., Tarksi, A. (eds.) The Axiomatic Method, with Special Reference to Geometry and Physics. Proceedings of an International Symposium held at the Univ. of Calif., Berkeley, December 26, 1957–January 4, 1958. pp. 16–29. Studies in Logic and the Foundations of Mathematics, North-Holland, Amsterdam (1959), available as a 2007 reprint, Brouwer Press, ISBN 1-443-72812-8
32. Tarski, A., Givant, S.: Tarski's system of geometry. The Bulletin of Symbolic Logic 5(2), 175–214 (1999)
33. Veblen, O.: A system of axioms for geometry. Transactions of the American Mathematical Society 5, 343–384 (1904)
34. Wos, L.: Automated reasoning and the discovery of missing and elegant proofs. Rinton Press, Paramus (2003)
35. Wu, W.T.: Mechanical Theorem Proving in Geometries: Basic Principles. Springer, Wien (1994)
36. Ziegler, M.: Einige unentscheidbare körpertheorien. Enseignement Math. 2(28), 269–280 (1982)

Automation of Geometry –
Theorem Proving, Diagram Generation,
and Knowledge Management

Dongming Wang

Laboratoire d'Informatique de Paris 6, Université Pierre et Marie Curie – CNRS,
4 Place Jussieu, 75252 Paris Cedex 05, France

Abstract of Talk

The process of theorem proving in geometry is sophisticated and intelligence-demanding. Mechanizing this process has been the objective of many great scientists, from ancient times to the information era. Scientific breakthroughs and technological advances have been made in the last three decades, which allows us now to automate the process (almost) fully on modern computing devices. The remarkable success of automated theorem proving has been a major source of stimulation for investigations on the automation of other processes of geometric deduction such as diagram generation and knowledge management.

This talk provides an account of historical developments on the mechanization and the automation of theorem proving in geometry, highlighting representative methodologies and approaches. Automated generation of dynamic diagrams involving both equality and inequality constraints is discussed as another typical task of geometric deduction. The presentation is then centered around the concept and the management of geometric knowledge. We view geometric theorems, proofs, and diagrams as well as methods as knowledge objects and thus as part of the geometric knowledge. We are interested in creating reliable software environments in which different kinds of geometric knowledge are integrated, effective algorithms and techniques for managing the knowledge are implemented, and the user can use the built-in knowledge data and functions to develop new tools and to explore geometry visually, interactively, and dynamically.

We have considered and studied several foundational and engineering issues of geometric knowledge management and adopted some key strategies to deal with the issues. We explain and discuss such issues and strategies and demonstrate the effectiveness of the strategies by some pieces of software that have implemented preliminary and experimental versions of our geometric knowledge base, geometric-object-oriented language, and geometric textbook system.

T. Ida and J. Fleuriot (Eds.): ADG 2012, LNAI 7993, pp. 31–32, 2013.

References

1. Bibliography on Geometric Reasoning (2003),
   http://www-polsys.lip6.fr/~wang/GRBib/
2. Chen, X., Li, W., Luo, J., Wang, D.: Open geometry textbook: A case study of knowledge acquisition via collective intelligence (project description). In: Jeuring, J., Campbell, J.A., Carette, J., Dos Reis, G., Sojka, P., Wenzel, M., Sorge, V. (eds.) CICM 2012. LNCS (LNAI), vol. 7362, pp. 432–437. Springer, Heidelberg (2012)
3. Chen, X., Wang, D.: Management of geometric knowledge in textbooks. Data & Knowledge Engineering 73, 43–57 (2012)
4. Liang, T., Wang, D.: On the design and implementation of a geometric-object-oriented language. Frontiers of Computer Science in China 1(2), 180–190 (2007)
5. Zhao, T., Wang, D., Hong, H., Aubry, P.: Real solution formulas of cubic and quartic equations applied to generate dynamic diagrams with inequality constraints. In: Proceedings of the 27th ACM Symposium on Applied Computing (SAC 2012), Riva del Garda, Italy, March 25–29, pp. 94–101. ACM Press, New York (2012)

Improving Angular Speed Uniformity
by C^1 Piecewise Reparameterization

Jing Yang[1], Dongming Wang[2], and Hoon Hong[3]

[1] LMIB – School of Mathematics and Systems Science, Beihang University,
Beijing 100191, China
yangjing@smss.buaa.edu.cn
[2] Laboratoire d'Informatique de Paris 6, CNRS – Université Pierre et Marie Curie,
4 Place Jussieu – BP 169, 75252 Paris cedex 05, France
Dongming.Wang@lip6.fr
[3] Department of Mathematics, North Carolina State University,
Box 8205, Raleigh, NC 27695, USA
hong@ncsu.edu

Abstract. We show how to compute a C^1 piecewise-rational reparameterization that closely approximates to the arc-angle parameterization of any plane curve by C^1 piecewise Möbius transformation. By making use of the information provided by the first derivative of the angular speed function, the unit interval is partitioned such that the obtained reparameterization has high uniformity and continuous angular speed. An iteration process is used to refine the interval partition. Experimental results are presented to show the performance of the proposed method and the geometric behavior of the computed reparameterizations.

Keywords: Parametric plane curve, angular speed uniformity, C^1 piecewise Möbius transformation, monotonic behavior.

1 Introduction

Parametric curves have been used extensively in areas such as computer aided geometric design, computer graphics, and computer vision. A curve may have infinitely many different parameterizations. Depending on where and how the curve will be used, it may be necessary to find a suitable or optimal parameterization out of the infinitely many or to convert a given parameterization into another (more) suitable one. In this paper, we are concerned with parameterizations used for plotting, so typical choices are arc-length [1, 2, 5–8, 10, 15], chord-length [3, 11, 13] and arc-angle [9, 12, 17, 18].

This is the third in a series of papers in which we study the problem of reparameterizing plane curves to improve their angular speed uniformity (or closeness to arc-angle parameterization). In the first paper [17], we proposed a method for finding the optimal reparameterization of any plane curve among those obtained by Möbius transformations. In the second paper [18], we allowed C^0 piecewise Möbius transformations. The computed C^0 piecewise-rational reparameterization can approximate to the uniform parameterization as closely as one wishes,

T. Ida and J. Fleuriot (Eds.): ADG 2012, LNAI 7993, pp. 33–47, 2013.
© Springer-Verlag Berlin Heidelberg 2013

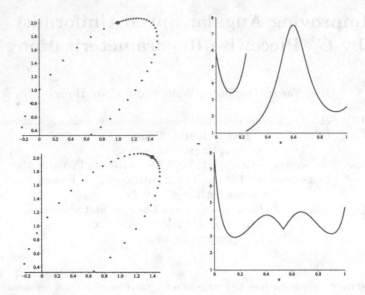

Fig. 1. Dot plots (left) and angular speeds (right) of two parameterizations of the same curve. The first row shows one parameterization with discontinuous angular speed and the second row shows the other with continuous angular speed. The red dots on the left-hand side correspond to the segment points.

but its angular speed sometimes lacks continuity, causing sudden changes in the density of points in plotting (see Fig. 1). In this third paper, we address the problem of discontinuity by restricting piecewise Möbius transformations to be C^1. We describe a method that can produce almost uniform reparameterizations with continuous angular speed.

In Section 2, we state the problem of C^1 piecewise reparameterization. In Section 3, we give an algorithm for computing a C^1 piecewise reparameterization from any given parameterization of a plane curve. In Section 4, we provide experimental results on the performance of the algorithm and compare them with the results obtained using C^0 piecewise Möbius transformations.

2   Problem

We begin by recalling some notions and results from [17, 18]. For any regular parameterization

$$p = (x(t), y(t)) : [0, 1] \to \mathbb{R}^2$$

of a plane curve, we denote its angle, angular speed, average angular speed and the L_2 norm of angular speed by $\theta_p, \omega_p, \mu_p$ and σ_p^2 respectively. They are defined by the following expressions

$$\theta_p = \arctan \frac{y'}{x'}, \quad \omega_p = |\theta_p'|, \quad \mu_p = \int_0^1 \omega_p(t)\, dt, \quad \sigma_p^2 = \int_0^1 (\omega_p(t) - \mu_p)^2\, dt.$$

$$(1)$$

Definition 1 (Angular Speed Uniformity). The *angular speed uniformity* u_p of p is defined as

$$u_p = \frac{1}{1 + \sigma_p^2/\mu_p^2} \tag{2}$$

(with $u_p = 1$ when $\mu_p = 0$).

Let
$$T = (t_0, \ldots, t_N), \quad S = (s_0 \ldots, s_N), \quad \alpha = (\alpha_0, \ldots, \alpha_{N-1}),$$
where $0 = t_0 < \cdots < t_N = 1$, $0 = s_0 < \cdots < s_N = 1$, and $0 < \alpha_0, \ldots, \alpha_{N-1} < 1$.

Definition 2 (C^1 Piecewise Möbius Transformation). A map m is called a C^1 *piecewise Möbius transformation* if it has the following form   .

$$m(s) = \begin{cases} \vdots \\ m_i(s), & \text{if } s \in [s_i, s_{i+1}]; \\ \vdots \end{cases} \tag{3}$$

such that
$$m_i'(s_{i+1}) = m_{i+1}'(s_{i+1}), \tag{4}$$

where
$$m_i(s) = t_i + \Delta t_i \frac{(1 - \alpha_i)\tilde{s}}{(1 - \alpha_i)\tilde{s} + (1 - \tilde{s})\alpha_i}$$
and $\Delta t_i = t_{i+1} - t_i$, $\Delta s_i = s_{i+1} - s_i$, $s = (s - s_i)/\Delta s_i$.

Now we are ready to state the problem precisely. Let p be a rational parameterization of a plane curve. The problem is essentially one of constrained optimization: find T, S and α such that $u_{p \circ m}$ is near maximum, subject to the constraint (4), where m is the transformation determined by T, S and α.

Remark 1. The constraint (4) ensures that the resulting reparameterization is C^1 continuous, since $\omega_{p \circ m} = (\omega_p \circ m)(s) \cdot m'(s)$.

Remark 2. From now on, we make the natural assumption that p is not a straight line and $\omega_p \neq 0$ for all $t \in [0, 1]$.

3   Method

Recall the problem of constrained optimization formulated in the preceding section. Finding an exact optimal solution to this problem is difficult both theoretically and practically due to the highly nonlinear dependency of $u_{p \circ m}$ on T, S, α. Therefore we will try to compute an approximately optimal solution instead.

3.1 Determination of T

It is proved in [17] that if a transformation r satisfies the condition

$$(r^{-1})' = \frac{\omega_p}{\mu_p}, \tag{5}$$

then $u_{por} = 1$, that is, $p \circ r$ is an arc-angle parameterization. Since μ_p is a constant, the condition (5) means that $(r^{-1})'$ is proportional to ω_p. Hence, if $(m_i^{-1})'$ is similar to ω_p over $[t_i, t_{i+1}]$, then m_i is close to the optimal r over $[t_i, t_{i+1}]$. In this case, $p \circ m$ will be a good reparameterization.

As given in Definition 2,

$$m_i(s) = t_i + \Delta t_i \frac{(1 - \alpha_i)\tilde{s}}{(1 - \alpha_i)\tilde{s} + (1 - \tilde{s})\alpha_i}.$$

Taking derivative of both sides of this equality with respect to s, we obtain

$$m_i'(s) = \frac{\Delta t_i}{\Delta s_i} \cdot \frac{\alpha_i(1 - \alpha_i)}{[(1 - \alpha_i)\tilde{s} + \alpha_i(1 - \tilde{s})]^2}. \tag{6}$$

It follows that

$$(m_i^{-1})'(t) = \frac{\Delta s_i}{\Delta t_i} \cdot \frac{(1 - \alpha_i)\,\alpha_i}{\left[(1 - \alpha_i)(1 - \tilde{t}) + \alpha_i \tilde{t}\right]^2} \tag{7}$$

(where the derivative is with respect to t, and similarly elsewhere). Observe that each $(m_i^{-1})'$ is monotonic over $[t_i, t_{i+1}]$. If ω_p is also monotonic over $[t_i, t_{i+1}]$, then ω_p and $(m_i^{-1})'$ are likely similar and approximately proportional. This suggests that $t_1, \ldots, t_{N-1}$ should be chosen such that

$$\omega_p'(t_i) = 0.$$

Example 1. Consider the parameterization

$$p = \left(\frac{t^3 - 6\,t^2 + 9\,t - 2}{2\,t^4 - 16\,t^3 + 40\,t^2 - 32\,t + 9}, \frac{t^2 - 4\,t + 4}{2\,t^4 - 16\,t^3 + 40\,t^2 - 32\,t + 9} \right),$$

whose plot is shown in Fig. 2. The angular speed of p is

$$\omega_p = \left| \frac{2(2\,t^4 - 16\,t^3 + 40\,t^2 - 32\,t + 9) \cdot G}{F} \right|,$$

where

$$F = 4\,t^{12} - 96\,t^{11} + 1032\,t^{10} - 6560\,t^9 + 27448\,t^8 - 79744\,t^7 + 165784\,t^6$$
$$- 251680\,t^5 + 283789\,t^4 - 239208\,t^3 + 144730\,t^2 - 53800\,t + 8753,$$
$$G = 2\,t^6 - 24\,t^5 + 138\,t^4 - 464\,t^3 + 939\,t^2 - 1068\,t + 551.$$

Now

$$\omega_p' = -\frac{4\,(t-2)\cdot H}{F^2},$$

where

$$
\begin{aligned}
H = {}& 16\,t^{20} - 640\,t^{19} + 12320\,t^{18} - 151680\,t^{17} + 1336544\,t^{16} - 8928256\,t^{15} \\
& + 46713024\,t^{14} - 195122432\,t^{13} + 657869152\,t^{12} - 1800130816\,t^{11} \\
& + 4002473088\,t^{10} - 7212888576\,t^9 + 10468965804\,t^8 - 12110797504\,t^7 \\
& + 10998339008\,t^6 - 7681616640\,t^5 + 4019329519\,t^4 - 1526783864\,t^3 \\
& + 406515286\,t^2 - 72301976\,t + 7081867.
\end{aligned}
$$

By solving $\omega_p'(t) = 0$ for t over $(0,1)$, we obtain $T \doteq (0, 0.580, 1)$.

3.2 Determination of S

Recall (5), which may be rewritten into

$$r^{-1} = \frac{\int_0^t \omega_p\, dt}{\mu_p}.$$

If m is a C^1 piecewise Möbius transformation which approximates to r closely, then $s = m^{-1} \doteq \int_0^t \omega_p\, dt/\mu_p$. This suggests us to choose $s_1, \ldots, s_{N-1}$ with

$$s_i = \frac{\int_0^{t_i} \omega_p\, dt}{\mu_p}. \tag{8}$$

Example 2 (Continued from Example 1). For the parameterization p given in Example 1, we have

$$\mu_p = \int_0^1 \omega_p\, dt \doteq 3.811, \qquad s_1 \doteq \frac{\int_0^{0.580} \omega_p\, dt}{3.811} \doteq 0.535.$$

Thus $S \doteq (0, 0.535, 1)$.

3.3 Determination of α

We want to solve the constraint (4) for $\alpha_1, \ldots, \alpha_{N-1}$ in terms of T, S and α_0. For this purpose, we use a few shorthand notations:

$$\lambda_i = \frac{\Delta t_i}{\Delta s_i}, \qquad \rho_i = \frac{1 - \alpha_i}{\alpha_i}, \tag{9}$$

$$\phi_{k+1} = \prod_{i=0}^k \lambda_{2i}, \qquad \psi_{k+1} = \prod_{i=0}^k \lambda_{2i+1}$$

with

$$\phi_0 = \psi_0 = 1.$$

The following Lemma first appeared in [2] without a proof.

Lemma 1. *Let* $\Psi_k = \dfrac{\psi_k^2}{\phi_k^2}$ *and* $\Phi_k = \dfrac{\phi_{k+1}^2}{\psi_k^2}$. *The constraint* (4) *is equivalent to*

$$
\rho_i = \begin{cases} \dfrac{\Psi_k \rho_0 \lambda_0}{\lambda_i}, & \text{if } i = 2\,k; \\[3mm] \dfrac{\Phi_k}{\rho_0 \lambda_0 \lambda_i}, & \text{if } i = 2\,k+1. \end{cases}
\tag{10}
$$

Proof. Refer to (6). Using the shorthand notations, we have

$$
m_i'(s) = \lambda_i \frac{\rho_i}{[(1-\tilde{s}) + \rho_i \tilde{s}]^2}.
$$

Hence the constraint (4) implies

$$
\frac{\lambda_i}{\rho_i} = \lambda_{i+1} \rho_{i+1}.
\tag{11}
$$

We prove the lemma by induction on i. When $i = 0$, $k = 0$ and the conclusion can be easily verified. Assume that the conclusion holds for a positive i. If $i = 2\,k$, then by (11)

$$
\rho_{i+1} = \frac{\lambda_i^2}{\lambda_i \rho_i} \cdot \frac{1}{\lambda_{i+1}} = \frac{\lambda_{2k}^2}{\Psi_k \rho_0 \lambda_0} \cdot \frac{1}{\lambda_{i+1}} = \frac{\lambda_{2k}^2}{\frac{\psi_k^2}{\phi_k^2} \cdot \frac{\rho_0 \lambda_0}{1}} \cdot \frac{1}{\lambda_{i+1}}
$$

$$
= \frac{\phi_k^2 \lambda_{2k}^2}{\psi_k^2} \cdot \frac{1}{\rho_0 \lambda_0} \cdot \frac{1}{\lambda_{i+1}} = \frac{\phi_{k+1}^2}{\psi_k^2} \cdot \frac{1}{\rho_0 \lambda_0 \lambda_{i+1}} = \frac{\Phi_k}{\rho_0 \lambda_0 \lambda_{i+1}}.
$$

Similarly, if $i = 2\,k+1$, then

$$
\rho_{i+1} = \frac{\lambda_i^2}{\lambda_i \rho_i} \cdot \frac{1}{\lambda_{i+1}} = \frac{\lambda_{2k+1}^2}{\frac{\Phi_k}{\rho_0 \lambda_0}} \cdot \frac{1}{\lambda_{i+1}} = \frac{\lambda_{2k+1}^2}{\frac{\phi_{k+1}^2}{\psi_k^2} \cdot \frac{1}{\rho_0 \lambda_0}} \cdot \frac{1}{\lambda_{i+1}}
$$

$$
= \frac{\psi_k^2 \lambda_{2k+1}^2}{\phi_{k+1}^2} \cdot \frac{\rho_0 \lambda_0}{1} \cdot \frac{1}{\lambda_{i+1}} = \frac{\psi_{k+1}^2}{\phi_{k+1}^2} \cdot \frac{\rho_0 \lambda_0}{\lambda_{i+1}} = \frac{\Psi_{k+1} \rho_0 \lambda_0}{\lambda_{i+1}}.
$$

Therefore the conclusion holds in both cases. □

Next, we will express the objective function u_{pom} in terms of T, S and α_0. For this end, we use the following shorthand notations:

$$
A_i = \int_{t_i}^{t_{i+1}} \omega_p^2(t) \cdot (1 - \tilde{t})^2 dt,
$$

$$
B_i = \int_{t_i}^{t_{i+1}} \omega_p^2(t) \cdot 2\,\tilde{t}(1 - \tilde{t})\, dt,
$$

$$
C_i = \int_{t_i}^{t_{i+1}} \omega_p^2(t) \cdot \tilde{t}^2 dt.
$$

Lemma 2. *The following equality holds for any parameterization p and C^1 piecewise Möbius transformation m:*

$$u_{pom} = \frac{\mu_p^2}{\eta_{p,m}},\qquad(12)$$

where

$$\eta_{p,m} = \sum_{k=0}^{\lfloor \frac{N-1}{2}\rfloor} \left[\Psi_k \rho_0\lambda_0 A_{2k} + \lambda_{2k}B_{2k} + \frac{\Phi_k}{\rho_0\lambda_0}C_{2k} \right.$$
$$\left. + \frac{\Phi_k}{\rho_0\lambda_0}A_{2k+1} + \lambda_{2k+1}B_{2k+1} + \Psi_{k+1}\rho_0\lambda_0 C_{2k+1} \right]$$

and those terms above whose subscripts are greater than $N-1$ take value 0.

Proof. It is proved in [17] that

$$u_{pom} = \frac{\mu_p^2}{\eta_{p,m}},\quad \text{where}\quad \eta_{p,m} = \int_0^1 \frac{\omega_p^2}{(m^{-1})'}(t)\,dt$$

and m^{-1} is the inverse function of m. Since m is piecewise, we have

$$\eta_{p,m} = \sum_{i=0}^{N-1} \int_{t_i}^{t_{i+1}} \frac{\omega_p^2(t)}{(m_i^{-1})'(t)}\,dt.$$

It follows from (7) that

$$\frac{1}{(m_i^{-1})'(t)} = \frac{\Delta t_i}{\Delta s_i} \cdot \frac{[(1-\alpha_i)(1-\tilde{t}) + \alpha_i\tilde{t}]^2}{(1-\alpha_i)\alpha_i}$$
$$= \lambda_i\left[\rho_i(1-\tilde{t})^2 + 2\tilde{t}(1-\tilde{t}) + \frac{1}{\rho_i}\tilde{t}^2\right],$$

where λ_i and ρ_i are as in (9). Hence

$$\eta_{p,m} = \sum_{i=0}^{N-1} \lambda_i\left(A_i\rho_i + B_i + \frac{C_i}{\rho_i}\right).\qquad(13)$$

The lemma is proved by substituting the expressions of ρ_i in Lemma 1 into (13). □

Theorem 1. *Let T and S be fixed. Then u_{pom} has a unique global maximum at*

$$\rho_0 = \frac{1}{\lambda_0}\sqrt{\frac{P}{Q}},\qquad(14)$$

where

$$P = \sum_{k=0}^{\lfloor \frac{N-1}{2}\rfloor} \Phi_k\left(A_{2k+1} + C_{2k}\right),\quad Q = \sum_{k=0}^{\lfloor \frac{N-1}{2}\rfloor}\left(\Psi_k A_{2k} + \Psi_{k+1}C_{2k+1}\right).\qquad(15)$$

Proof. As μ_p is a constant, we only need to minimize $\eta_{p,m}$. It is easy to verify that $\eta_{p,m} \geq \mu_p^2$. From (13), we see that when ρ_0 approaches 0 or ∞ the value of $\eta_{p,m}$ goes to $+\infty$. Thus there must be a global minimum point $\rho_0 \in (0, \infty)$ such that

$$\frac{d\eta_{p,m}}{d\rho_0} = 0.$$

Rewrite $\frac{d\eta_{p,m}}{d\rho_0} = 0$ into

$$-Q \cdot \lambda_0 + P \frac{1}{\rho_0^2 \lambda_0} = 0,$$

where P and Q are as in (15). Solving the above equation for ρ_0, we obtain

$$\rho_0 = \frac{1}{\lambda_0} \sqrt{\frac{P}{Q}}.$$

$\square$

Example 3 (Continued from Example 2). For p and T in Example 1 and S in Example 2, calculations result in $P \doteq 35.864$, $Q \doteq 0.740$ and $\lambda_0 \doteq 1.086$, so $\rho_0 \doteq 6.414$. By (11), we have $\rho_1 \doteq 0.188$ and thus $\alpha \doteq (0.135, 0.842)$.

3.4   Iteration

Once T, S and α are computed, we can construct a C^1 piecewise Möbius transformation m_1 for p such that $p_1 = p \circ m_1$ has better uniformity than p. Then we can use the same method to construct another C^1 piecewise Möbius transformation m_2 for p_1 such that $p_2 = p_1 \circ m_2$ has better uniformity than p_1, and repeat the process. In this way, we may get $m_1, \ldots, m_n$ and $p_1, \ldots, p_n$ such that $p_n = p_{n-1} \circ m_n = p \circ m_1 \circ \cdots \circ m_n$ has the desired uniformity.

The approach explained above involves extensive computations with floating-point numbers, resulting sometimes in instability, in particular in the process of solving the equation $\omega'_{p \circ m_1 \circ \cdots \circ m_k} = 0$. In what follows we propose an alternative approach that can avoid manipulating floating-point numbers.

Note that if m_1 and m_2 are C^1 piecewise Möbius transformations, then so is $m_1 \circ m_2$. Therefore finding $m_1, \ldots, m_n$ such that p_n has the desired uniformity is equivalent to finding a single m such that $m = m_1 \circ \cdots \circ m_n$ and $p \circ m$ has the desired uniformity. Thus the key question is how to find the partition for $m_1 \circ \cdots \circ m_n$ without computing each m_i.

Suppose that T and α have been computed by using the method presented in Sections 3.1–3.3. Let $t_i, t_{i+1} \in T$ and $\alpha_i \in \alpha$. Now we compute $t_{ij} \in (t_i, t_{i+1})$ such that the corresponding $s_{ij} \in (s_i, s_{i+1})$ are the partition nodes satisfying $\omega'_{p \circ m}(s_{ij}) = 0$. According to [17], we have

$$\omega_{p \circ m} = \frac{\omega_p}{(m^{-1})'}(t).$$

It follows that

$$(\omega_{p \circ m})' = \left[\frac{\omega_p}{(m^{-1})'} \right]' = \frac{\omega_p'(m^{-1})' - \omega_p(m^{-1})''}{[(m^{-1})']^2}.$$

Combining (10) and (14), we have $\rho_i \in (0, +\infty)$, which implies $\alpha_i \in (0, 1)$. By (7), $(m^{-1})' > 0$ when $t \in (t_i, t_{i+1})$. Thus the solution to $(\omega_{pom})' = 0$ for t over (t_i, t_{i+1}) is the same as that to

$$\omega_p'(m^{-1})' - \omega_p(m^{-1})'' = 0. \tag{16}$$

Substituting (7) and

$$(m^{-1})'' = \frac{\Delta s_i}{\Delta t_i^2} \cdot \frac{2\,\alpha_i(1 - \alpha_i)(1 - 2\,\alpha_i)}{\left[\alpha_i \tilde{t} + (1 - \alpha_i)(1 - \tilde{t})\right]^3}$$

into (16) and simplifying the result, we obtain

$$\omega_p' \cdot \Delta t_i \cdot \left[\alpha_i \tilde{t} + (1 - \alpha_i)(1 - \tilde{t})\right] - 2\,\omega_p \cdot (1 - 2\,\alpha_i) = 0. \tag{17}$$

By solving this equation for t, we can get a partition T_i of (t_i, t_{i+1}).

From T_i $(i = 0, \ldots, N-1)$, we can compute a refined partition T^* of $[0, 1]$. The partition T^* may be further refined in the same way by iteration. From T^* and p, we may compute a reparameterization $p \circ m$ of p, which is equivalent to $p \circ m_1 \circ \cdots \circ m_n$.

Example 4 (Continued from Example 3). Refer to Examples 1 and 3 for ω_p, ω_p', T and α. Consider the first interval, i.e., $(0, 0.580)$, in the partition T. To compute a partition of this interval, we substitute $t_0 = 0$, $t_1 \doteq 0.580$, $\alpha_0 \doteq 0.135$, ω_p and ω_p' into (17). The solution to the obtained equation gives a partition of $(0, 0.580)$, which is $T_1 \doteq (0, 0.424, 0.553, 0.580)$. Similarly for $(0.580, 1)$, we can also get a partition of it, which is $T_2 \doteq (0.580, 0.611, 0.749, 1)$. Thus

$$T^* \doteq (0, 0.424, 0.553, 0.580, 0.611, 0.749, 1)$$

is a refined partition of $[0, 1]$ we wanted.

Remark 3. The above alternative approach allows one to compute a refined partition of the unit interval by solving equations of the form (17). Unlike the bruteforce approach of iteration, this approach computes ω_p and ω_p' only once. Moreover, the integrations A_i, B_i, C_i are computed also much more effectively, using this approach than using the bruteforce one, because the latter involves integrations of piecewise functions many times, which usually take a large amount of computing time.

Remark 4. In practice, with two or three iterations one may reach the desired uniformity, e.g., $u \geq 0.99$.

3.5 Algorithm

Now we summarize the ideas and results discussed above into an algorithm.

Algorithm 1 (C^1_Reparameterize)
Input: p, a rational parameterization of a plane curve;
           δ, a real number greater than 1.

Output: p^*, a C^1 piecewise-rational reparameterization of p;
           u^*, the uniformity of p^*.

1.   Compute ω_p and μ_p using (1), u_p using (2) and ω'_p.

2. $m_{\text{new}} \leftarrow \text{Id}$; $u_{\text{new}} \leftarrow u_p$.

3.   Do

        $m_{\text{old}} \leftarrow m_{\text{new}}$;

        $u_{\text{old}} \leftarrow u_{\text{new}}$;

        $m_{\text{new}}, u_{\text{new}} \leftarrow \text{Improve}(m_{\text{old}}, \omega_p, \omega'_p, \mu_p)$

    Until $u_{\text{new}}/u_{\text{old}} < \delta$.

4. $p^* \leftarrow p \circ m_{\text{new}}$; $u^* \leftarrow u_{\text{new}}$.

5.   Return p^*, u^*.

Algorithm 2 (Improve)
Input: m_{old}, a C^1 piecewise Möbius transformation;

           ω_p, the angular speed of a parametric plane curve p;

           ω'_p, the derivative of ω_p;

           μ_p, the average angular speed of p.

Output: m_{new}, an improved C^1 piecewise Möbius transformation for p;
           u_{new}, the uniformity of $p \circ m_{\text{new}}$.

1.   Let $T_{\text{old}}, S_{\text{old}}$ and α_{old} be the parameters for m_{old}.

2.   Compute T_{new} from T_{old} and α_{old}, using (17).

3.   Compute S_{new} from T_{new}, using (8).

4.   Compute $\rho_{0\text{new}}$ from T_{new} and S_{new}, using (14).

5.   Compute $\rho_{i\text{new}}$ $(i > 0)$ from $T_{\text{new}}, S_{\text{new}}$ and $\rho_{0\text{new}}$, using (10).

6.   Compute α_{new} from $\rho_{i\text{new}}$, using (9).

7.   Compute u_{new} from μ_p and $\rho_{i\text{new}}$, using (12) and (13).

8.   Construct m_{new} from $T_{\text{new}}, S_{\text{new}}$ and α_{new}, using (3).

9.   Return $m_{\text{new}}, u_{\text{new}}$.

Remark 5. In the main algorithm C^1_Reparameterize, ω_p, ω'_p and μ_p are computed only once and used repeatedly in the iteration step as input to the sub-algorithm Improve. The iteration terminates when the condition $u_{\text{new}}/u_{\text{old}} < \delta$ is satisfied, i.e., when it cannot improve the uniformity by a factor greater than or equal to the given δ.

3.6 Example

Example 5 (Continued from Example 4). For p in Example 1 with $\delta = 1.01$, the algorithm C^1-Reparameterize starts with the computation of ω_p, μ_p and ω'_p (see Examples 1 and 2 for the results). After setting $m_{\mathsf{old}}(s) = s$ and $u_{\mathsf{old}} = 0.403$, the algorithm calls Improve with ω_p, ω'_p, μ_p and m_{old} as input. With $T_{\mathsf{old}} = (0, 1)$, $S_{\mathsf{old}} = (0, 1)$ and $\alpha_{\mathsf{old}} = 1/2$ as the parameters for m_{old}, we compute T_{new}, S_{new}, α_{new} (as done in Examples 1–3) and $u_{\mathsf{new}} \doteq 0.969$, and get

$$
m_{\mathsf{new}}(s) \doteq
\begin{cases}
\dfrac{0.939\,s}{1.366\,s + 0.135}, & s \in [0, 0.535]; \\[3mm]
\dfrac{0.710\,s - 0.869}{1.469\,s - 1.628}, & s \in [0.535, 1].
\end{cases}
$$

Since $u_{\mathsf{new}}/u_{\mathsf{old}} \doteq 2.406 > \delta$, we update m_{old} with m_{new} and u_{old} with u_{new} and call Improve. Now with $T_{\mathsf{old}} \doteq (0, 0.580, 1)$, $S_{\mathsf{old}} \doteq (0, 0.535, 1)$ and $\alpha_{\mathsf{old}} \doteq (0.135, 0.842)$ as the parameters for m_{old}, we compute

$$
\begin{aligned}
T_{\mathsf{new}} &\doteq (0, 0.424, 0.553, 0.580, 0.611, 0.749, 1), \\
S_{\mathsf{new}} &\doteq (0, 0.170, 0.402, 0.535, 0.679, 0.905, 1), \\
\alpha_{\mathsf{new}} &\doteq (0.359, 0.284, 0.481, 0.523, 0.729, 0.617), \\
u_{\mathsf{new}} &\doteq 0.997,
\end{aligned}
$$

and get

$$
m_{\mathsf{new}}(s) \doteq
\begin{cases}
\dfrac{1.601\,s}{1.657\,s + 0.359}, & s \in [0, 0.170]; \\[3mm]
\dfrac{1.185\,s - 0.080}{1.855\,s + 0.030}, & s \in [0.170, 0.402]; \\[3mm]
\dfrac{0.264\,s + 0.160}{0.285\,s + 0.367}, & s \in [0.402, 0.535]; \\[3mm]
\dfrac{0.084\,s - 0.348}{0.315\,s - 0.691}, & s \in [0.535, 0.679]; \\[3mm]
\dfrac{1.068\,s - 1.170}{2.022\,s - 2.101}, & s \in [0.679, 0.905]; \\[3mm]
\dfrac{0.829\,s - 1.212}{2.461\,s - 2.844}, & s \in [0.905, 1].
\end{cases}
$$

Since $u_{\mathsf{new}}/u_{\mathsf{old}} \doteq 1.029$, which is still bigger than δ, we update m_{old} with m_{new} and u_{old} with u_{new}, and call Improve again. With the updated T_{old}, S_{old} and α_{old}

as the parameters for m_{old}, we compute

$$T_{new} \doteq (0, 0.323, 0.424, 0.476, 0.523, 0.553, 0.572, 0.580, 0.589,$$
$$0.611, 0.648, 0.683, 0.749, 0.839, 1),$$
$$S_{new} \doteq (0, 0.113, 0.170, 0.220, 0.302, 0.402, 0.492, 0.535, 0.582,$$
$$0.679, 0.790, 0.847, 0.905, 0.950, 1),$$
$$\alpha_{new} \doteq (0.446, 0.433, 0.436, 0.411, 0.435, 0.475, 0.499, 0.501,$$
$$0.529, 0.582, 0.568, 0.581, 0.563, 0.552),$$
$$u_{new} \doteq 1.000,$$

and get

$$m_{new}(s) \doteq \begin{cases} \dfrac{1.589\,s}{0.953\,s + 0.446}, & s \in [0, 0.113]; \\[2ex] \dfrac{1.760\,s - 0.058}{2.345\,s + 0.169}, & s \in [0.113, 0.170]; \\[1ex] \quad\vdots \\[1ex] \dfrac{0.321\,s - 0.768}{2.094\,s - 2.542}, & s \in [0.950, 1] \end{cases}$$

(which consists of 14 pieces).

Since $u_{new}/u_{old} \doteq 1.003 < \delta$, the iteration terminates. Finally, $p^* = p \circ m_{new}$ is obtained as a desired reparameterization with $u_{p^*} \doteq 1.000$. Figure 2 shows the original parameterization p and the improved reparameterization p^*.

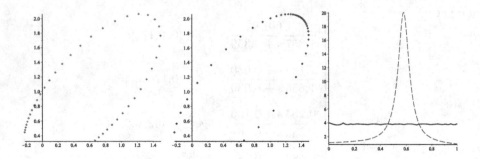

Fig. 2. Curve p and its reparameterization p^* computed by using a C^1 piecewise Möbius transformation. The dot plots using p and p^* are shown on the left-hand side and in the middle respectively. The dash and solid curves on the right-hand side show the angular speeds of p and p^* respectively.

Remark 6. In Examples 1–5 and the experiments in Section 4, T, S, α and m are all computed numerically. It is worth pointing out that if floating-point numbers are turned finally into rational numbers and $q = p \circ m$ is computed symbolically, then q is an exact reparameterization of p. However, the continuity of angular speed of q is not guaranteed to be exact unless ρ_i are computed also symbolically at the last stage.

4   Experiments

The algorithm described in Section 3.5 has been implemented in Maple. It performs well as shown by the experimental results given in this section. The experiments were carried out on a PC Intel(R) Core(TM)2 Duo CPU P8400 @2.26GHz with 2 GB of RAM. The benchmark curves, chosen from [4, 14, 16], are all rational and their angular speeds are nonzero over $[0, 1]$. For the curves taken from [4], the involved parameters are specialized to concrete values. The list of the curves (in Maple format) is available from the authors upon request.

Table 1. Reparameterization by standard, C^0 and C^1 piecewise Möbius transformations (with $\delta = 1.01$). In the table, u = Uniformity, N = Number of pieces, $\bar{D}$ = Discontinuity (average), T = Time (seconds).

Curve	Original	Standard		C^0 piecewise				C^1 piecewise		
	u	u	T	u	$\bar{D}$	N	T	u	N	T
C1	0.403	0.412	0.015	1.000	0.008	14	0.391	1.000	14	0.297
C2	0.552	0.558	0.000	1.000	0.023	6	0.047	1.000	6	0.062
C3	0.403	0.643	0.000	1.000	0.004	18	0.328	1.000	12	0.328
C4	0.926	0.973	0.000	1.000	0.013	4	0.047	1.000	4	0.062
C5	0.987	0.987	0.015	1.000	0.004	4	1.016	1.000	4	0.734
C6	0.797	0.797	0.000	1.000	0.004	22	0.359	1.000	18	0.407
C7	0.836	0.853	0.015	1.000	0.017	5	0.047	1.000	5	0.047
C8	0.741	0.741	0.016	1.000	0.015	8	0.093	0.999	8	0.110
C9	0.337	0.337	0.015	1.000	0.010	16	0.172	0.999	16	0.188
C10	0.347	0.347	0.015	1.000	0.008	16	0.219	1.000	16	0.235
C11	0.555	0.555	0.015	1.000	0.019	18	0.203	1.000	22	0.516
C12	0.764	0.764	0.015	1.000	0.006	22	0.328	1.000	18	0.328
C13	0.747	0.747	0.000	1.000	0.017	10	0.094	1.000	10	0.094
C14	0.301	0.301	0.015	0.999	0.010	4	0.031	0.999	4	0.047
C15	0.276	0.667	0.015	1.000	0.021	18	0.219	0.999	16	0.250
C16	0.283	0.283	0.000	1.000	0.010	16	0.203	0.999	16	0.218
C17	0.682	0.890	0.000	1.000	0.002	8	0.063	1.000	7	0.062
C18	0.226	0.228	0.000	1.000	0.000	8	0.078	1.000	8	0.078

Table 1 presents the experimental results of reparameterization using standard Möbius transformations, C^0 piecewise Möbius transformations and C^1 piecewise Möbius transformations. In order to make comparisons fair, the way used for computing partitions for C^0 piecewise reparameterizations here is different from that described in [18], but similar to the process of computing partitions for C^1 piecewise reparameterizations explained in Section 3.4. One can see that the uniformities of parameterizations may be improved only slightly by standard Möbius transformations, but dramatically by piecewise ones. C^0 piecewise reparameterization often introduces discontinuity in angular speed. For any piecewise parameterization p with $(0, t_1, \ldots, t_{N-1}, 1)$ as a partition of $[0, 1]$, the local discontinuity of the angular speed of p at t_i may be naturally defined to be

$$D_i = 2 \left| \frac{\omega_p(t_i^+) - \omega_p(t_i^-)}{\omega_p(t_i^+) + \omega_p(t_i^-)} \right|.$$

The total and average discontinuities of the angular speed of p may be measured by $D = \sum_{i=1}^{N-1} D_i$ and $\bar{D} = D/(N - 1)$ respectively. The total discontinuity of the angular speed of any C^1 piecewise parameterization is 0 (or almost 0 when computations are performed numerically). There is not much difference between the uniformities and computing times for C^0 and C^1 reparameterizations, while C^0 reparameterizations may have positive discontinuity, so we advocate the use of C^1 piecewise reparameterizations in practice.

Remark 7. In step 2 of Algorithm 2, all the solutions of the equation (17) are needed. We have encountered two test examples (not included in Table 1) for which the numeric solver of Maple used in our implementation failed to find all the solutions. For these two examples, the uniformities of the computed reparameterizations are 0.959 and 0.737 (of which the latter is not close to 1 at all).

5   Conclusion

We have presented a method for computing C^1 piecewise-rational reparameterizations of given parametric plane curves. Such reparameterizations may have significantly better uniformity and meanwhile keep their angular speed function continuous. The latter represents a major advantage of C^1 piecewise-rational reparameterizations over C^0 ones. In addition to the basic idea of computing C^1 piecewise Möbius transformations, the method contains two key ingredients for partitioning the unit interval: (1) the interval is first partitioned by using the information provided by the first derivative of the angular speed function; (2) the obtained interval partition is further refined by using an iteration process. Experimental results are provided to show the good performance of the proposed algorithm. It can be concluded that C^1 piecewise-rational reparameterizations computed by our algorithm may have uniformity almost as high as C^0 ones', but the former possess better geometric continuity and thus are more suitable for applications. The iteration idea used in this paper for interval partitioning may be used to compute C^1 piecewise arc-length reparameterizations as well, and thus to enhance the method described in [10].

Acknowledgements. This work has been supported partially by the Open Fund of SKLSDE under Grant No. SKLSDE-2011KF-02 and the ANR-NSFC project EXACTA (ANR-09-BLAN-0371-01/60911130369).

References

1. Cattiaux-Huillard, I., Albrecht, G., Hernández-Mederos, V.: Optimal parameterization of rational quadratic curves. Computer Aided Geometric Design 26(7), 725–732 (2009)
2. Costantini, P., Farouki, R., Manni, C., Sestini, A.: Computation of optimal composite re-parameterizations. Computer Aided Geometric Design 18(9), 875–897 (2001)
3. Farin, G.: Rational quadratic circles are parameterized by chord length. Computer Aided Geometric Design 23(9), 722–724 (2006)
4. Famous curves index (2005),
   http://www-history.mcs.st-and.ac.uk/Curves/Curves.html
5. Farouki, R.: Optimal parameterizations. Computer Aided Geometric Design 14(2), 153–168 (1997)
6. Farouki, R., Sakkalis, T.: Real rational curves are not unit speed. Computer Aided Geometric Design 8(2), 151–157 (1991)
7. Gil, J., Keren, D.: New approach to the arc length parameterization problem. In: Straßer, W. (ed.) Proceedings of the 13th Spring Conference on Computer Graphics, Budmerice, Slovakia, June 5–8, pp. 27–34. Comenius University, Slovakia (1997)
8. Jüttler, B.: A vegetarian approach to optimal parameterizations. Computer Aided Geometric Design 14(9), 887–890 (1997)
9. Kosters, M.: Curvature-dependent parameterization of curves and surfaces. Computer-Aided Design 23(8), 569–578 (1991)
10. Liang, X., Zhang, C., Zhong, L., Liu, Y.: C^1 continuous rational reparameterization using monotonic parametric speed partition. In: Proceedings of the 9th International Conference on Computer-Aided Design and Computer Graphics, Hong Kong, China, December 7–10, pp. 16–21. IEEE Computer Society, Hong Kong (2005)
11. Lü, W.: Curves with chord length parameterization. Computer Aided Geometric Design 26(3), 342–350 (2009)
12. Patterson, R., Bajaj, C.: Curvature adjusted parameterization of curves. Computer Science Technical Report CSD-TR-907, Paper 773, Purdue University, USA (1989)
13. Sánchez-Reyes, J., Fernández-Jambrina, L.: Curves with rational chord-length parametrization. Computer Aided Geometric Design 25(4–5), 205–213 (2008)
14. Sendra, J.R., Winkler, F., Pérez-Díaz, S.: Rational Algebraic Curves: A Computer Algebra Approach. Algorithms and Computation in Mathematics, vol. 22. Springer, Heidelberg (2008)
15. Walter, M., Fournier, A.: Approximate arc length parameterization. In: Velho, L., Albuquerque, A., Lotufo, R. (eds.) Proceedings of the 9th Brazilian Symposiun on Computer Graphics and Image Processing, Fortaleza-CE, Brazil, October 29–November 1, pp. 143–150. Caxambu, SBC/UFMG (1996)
16. Wang, D. (ed.): Selected Lectures in Symbolic Computation. Tsinghua University Press, Beijing (2003) (in Chinese)
17. Yang, J., Wang, D., Hong, H.: Improving angular speed uniformity by reparameterization. Computer Aided Geometric Design 30(7), 636–652 (2013)
18. Yang, J., Wang, D., Hong, H.: Improving angular speed uniformity by optimal C^0 piecewise reparameterization. In: Gerdt, V.P., Koepf, W., Mayr, E.W., Vorozhtsov, E.V. (eds.) CASC 2012. LNCS, vol. 7442, pp. 349–360. Springer, Heidelberg (2012)

Extending the Descartes Circle Theorem
for Steiner n-Cycles*

Shuichi Moritsugu

University of Tsukuba,
Tsukuba 305-8550, Ibaraki, Japan
moritsug@slis.tsukuba.ac.jp

Abstract. This paper describes the extension of the Descartes circle
theorem for Steiner n-cycles. Instead of using the inversion method, we
computed the Gröbner bases or resultants for the equations of inscribed
or circumscribed circles. As a result, we deduced several relations that
could be called the Descartes circle theorem for $n \geq 4$. We succeeded in
computing the defining polynomials of circumradii with degrees 4, 24,
and 48, for $n = 4, 5$, and 6, respectively.

1 Introduction

Given a circle O with radius R, and a circle I with radius r in the interior of O,
if there exists a closed chain of n mutually tangent circles with radii $r_1, \ldots, r_n$
between $O(R)$ and $I(r)$, it is called a Steiner n-cycle or a Steiner chain. For
example, Fig. 1 shows the case for $n = 4$. The purpose of this paper is to find
unknown identities among radii R, r, and r_i's for a Steiner n-cycle when $n \geq 4$,
extending the well-known Descartes circle theorem for $n = 3$. In this context,
we consider the following problem.

Problem 1. *Given the radii $r_1, r_2, \ldots, r_n$ in a Steiner n-cycle, compute the
radii R and r, or compute the relations among R, r, and r_i's.*

Classical proofs are applied to the formulae for Steiner chains using the inversion
of circles [2][3][4]. In contrast, we tried to find new formulae for a Steiner n-cycle
($n \geq 4$) by coordinates computation, as an alternative to the inversion method.
We therefore applied the Gröbner basis or resultant methods to the equations of
circles that touch mutually, and succeeded in computing the defining polynomial
$\varphi_n(R, r_i)$ of radius R for $n = 4, 5, 6$.

Other approaches of extension to m-dimensional Euclidean spaces [10][6] have
been studied, such as Soddy's hexlet in a 3-dimensional case. However, it seems
that formulae for $n \geq 4$ in a 2-dimensional plane have been seldom discussed to
date. For example, the extension by Wilker [13] forms a different figure from a
Steiner chain. Hence, we believe that our results contain some relations that are
unknown so far.

---

* This work was supported by a Grant-in-Aid for Scientific Research (22500004) from
the Japan Society for the Promotion of Science (JSPS).

T. Ida and J. Fleuriot (Eds.): ADG 2012, LNAI 7993, pp. 48–58, 2013.
© Springer-Verlag Berlin Heidelberg 2013

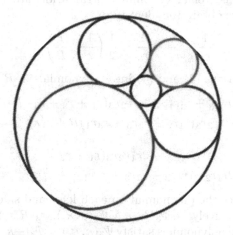

Fig. 1. A Steiner n-cycle ($n = 4$)

2   Previously Known Results

2.1   The Descartes Circle Theorem

Problem 1 for the case $n = 3$ was solved by Apollonius of Perga in the third century B.C., and the following formulae are now called the Descartes circle theorem, based on Descartes' letter to Princess Elizabeth of Bohemia written in 1643:

$$2\left(\frac{1}{R^2} + \frac{1}{r_1^2} + \frac{1}{r_2^2} + \frac{1}{r_3^2}\right) = \left(\frac{1}{r_1} + \frac{1}{r_2} + \frac{1}{r_3} - \frac{1}{R}\right)^2, \tag{1}$$

$$2\left(\frac{1}{r^2} + \frac{1}{r_1^2} + \frac{1}{r_2^2} + \frac{1}{r_3^2}\right) = \left(\frac{1}{r_1} + \frac{1}{r_2} + \frac{1}{r_3} + \frac{1}{r}\right)^2. \tag{2}$$

In these equations, we let the reciprocal of each radius be the curvature of each circle; that is, we put $K = 1/R$, $\kappa = 1/r$, $\kappa_1 = 1/r_1$, $\kappa_2 = 1/r_2$, and $\kappa_3 = 1/r_3$. Then, Equations (1) and (2) are expressed as follows:

$$K^2 + 2\left(\kappa_1 + \kappa_2 + \kappa_3\right)K + \kappa_1^2 + \kappa_2^2 + \kappa_3^2 - 2\left(\kappa_1\kappa_2 + \kappa_2\kappa_3 + \kappa_3\kappa_1\right) = 0, \tag{3}$$

$$\kappa^2 - 2\left(\kappa_1 + \kappa_2 + \kappa_3\right)\kappa + \kappa_1^2 + \kappa_2^2 + \kappa_3^2 - 2\left(\kappa_1\kappa_2 + \kappa_2\kappa_3 + \kappa_3\kappa_1\right) = 0. \tag{4}$$

In the setting of this paper, we note that $\kappa_1, \kappa_2, \kappa_3$ are given and K, κ are unknown. If we solve Equations (3) and (4) with K, κ and adopt suitable solutions, we obtain

$$K = -\left(\kappa_1 + \kappa_2 + \kappa_3\right) + 2\sqrt{\kappa_1\kappa_2 + \kappa_2\kappa_3 + \kappa_3\kappa_1}, \tag{5}$$

$$\kappa = \left(\kappa_1 + \kappa_2 + \kappa_3\right) + 2\sqrt{\kappa_1\kappa_2 + \kappa_2\kappa_3 + \kappa_3\kappa_1}, \tag{6}$$

which are referred to as Steiner's formulae. Straightforwardly, computing $\kappa - K = 2(\kappa_1 + \kappa_2 + \kappa_3)$, we obtain the identity

$$\frac{1}{r_1} + \frac{1}{r_2} + \frac{1}{r_3} = \frac{1}{2}\left(\frac{1}{r} - \frac{1}{R}\right). \tag{7}$$

If we transform Equations (3) and (4) into polynomials in R and r, we have

$$(r_1^2 r_2^2 + r_2^2 r_3^2 + r_3^2 r_1^2 - 2r_1 r_2 r_3(r_1 + r_2 + r_3))R^2$$
$$+2r_1 r_2 r_3(r_1 r_2 + r_2 r_3 + r_3 r_1)R + r_1^2 r_2^2 r_3^2 = 0, \tag{8}$$

$$(r_1^2 r_2^2 + r_2^2 r_3^2 + r_3^2 r_1^2 - 2r_1 r_2 r_3(r_1 + r_2 + r_3))r^2$$
$$-2r_1 r_2 r_3(r_1 r_2 + r_2 r_3 + r_3 r_1)r + r_1^2 r_2^2 r_3^2 = 0. \tag{9}$$

In the following, we let the polynomial on each left-hand side of Equations (3), (4), (8), and (9), respectively, be $\Phi_3(K, \kappa_i)$, $\Psi_3(\kappa, \kappa_i)$, $\varphi_3(R, r_i)$, and $\psi_3(r, r_i)$. We should note that these polynomials satisfy $\Psi_3(\kappa, \kappa_i) = \Phi_3(-\kappa, \kappa_i)$ and $\psi_3(r, r_i) = \varphi_3(-r, r_i)$.

Therefore, the objective of this paper is to compute and clarify the polynomials $\varphi_n(R, r_1, \ldots, r_n)$ and $\Phi_n(K, \kappa_1, \ldots, \kappa_n)$ for $n \geq 4$, using the relations among coordinates and radii of the circles.

2.2 Steiner's Porism

We assume that a circle $O(R)$ and a circle $I(r)$ in its interior are given and that d is the distance between their centers. If there exists a closed chain of n mutually tangent circles that form a Steiner n-cycle, we have the following relation:

$$(R - r)^2 - 4Rr \tan^2 \frac{\pi}{n} = d^2. \tag{10}$$

It is well known that if this closed chain exists once, it will always occur, irrespective of the position of the first circle of the chain. For example, Equation (10) is written as follows for $n = 3$ and 4, respectively:

$$(R - r)^2 - 12Rr = d^2, \tag{11}$$
$$(R - r)^2 - 4Rr = d^2. \tag{12}$$

For the case $n = 4$, it seems unknown whether any identities correspondent to the Descartes circle theorem shown in Equations (1) and (2) exist. Instead, the following relation that corresponds to Equation (7) is well known:

$$\frac{1}{r_1} + \frac{1}{r_3} = \frac{1}{r_2} + \frac{1}{r_4} = \frac{1}{r} - \frac{1}{R}. \tag{13}$$

Moreover, this can be extended to cases for a general even number $n(= 2k)$,

$$\frac{1}{r_1} + \frac{1}{r_{k+1}} = \frac{1}{r_2} + \frac{1}{r_{k+2}} = \cdots = \frac{1}{r_k} + \frac{1}{r_{2k}} = \frac{4(R - r)}{(R - r)^2 - d^2}. \tag{14}$$

When $n = 4$, if we apply Equation (12) to the above, Equation (13) is obtained. When $n(\geq 5)$ is odd, relations equivalent to Equation (7) seem to be unknown.

3 Computation by Gröbner Basis

3.1 Formalization

We now try to derive the identities described in the preceding section, by computing the Gröbner basis from the equations of circles and their tangent conditions [1]. First, we assume that circles $O(R) : x^2 + y^2 = R^2$ and $I(r) : x^2 + (y-d)^2 = r^2$ are given. Next, we draw the inner circles $x_i^2 + y_i^2 = r_i^2$ ($i = 1, \ldots, n$) touching one another successively. The circumscribing or inscribing relation is translated into the distance between the centers of each pair of circles:

$$f_i := x_i^2 + y_i^2 - (R - r_i)^2, \tag{15}$$

$$g_i := x_i^2 + (y_i - d)^2 - (r + r_i)^2, \tag{16}$$

$$h_i := (x_{i+1} - x_i)^2 + (y_{i+1} - y_i)^2 - (r_{i+1} + r_i)^2, \tag{17}$$

for $i = 1, \ldots, n$, where we read $x_{n+1} = x_1$, $y_{n+1} = y_1$, and $r_{n+1} = r_1$. Note that these polynomial representations contain both cases where the two circles are tangent internally and externally.

3.2 Computation for $n = 3$

We computed the Gröbner basis for the ideal $\{f_i, g_i, h_i\}$ using Maple 14, applying "loxdeg" group ordering in the Maple Gröbner package so that the variables were eliminated in the appropriate order. We note that Equations (1) and (2) expressing the Descartes circle theorem were independently computed. Hence, we consider the following three polynomial ideals using each monomial ordering:

(i) $J_R = (f_1, f_2, f_3, h_1, h_2, h_3)$, $[x_1, x_2, x_3, y_1, y_2, y_3] \succ [R, r_1, r_2, r_3]$
   The Gröbner basis in this case contains a polynomial $\varphi_3(R, r_1, r_2, r_3)$ that appears in Equation (8).
(ii) $J_r = (g_1, g_2, g_3, h_1, h_2, h_3)$, $[x_1, x_2, x_3, y_1, y_2, y_3, d] \succ [r, r_1, r_2, r_3]$
   The Gröbner basis in this case contains a polynomial $\psi_3(r, r_1, r_2, r_3)$ that appears in Equation (9).
(iii) $J = (f_1, f_2, f_3, g_1, g_2, g_3, h_1, h_2, h_3)$, $[x_1, x_2, x_3, y_1, y_2, y_3, r_1, r_2, r_3] \succ [r, R, d]$
   Here, the Gröbner basis contains a polynomial that is divisible by $R^2 - 14Rr + r^2 - d^2$ in Equation (11).

These results show that the Descartes circle theorem and Steiner's porism for $n = 3$ are confirmed by Gröbner basis computation. Here, we have $\psi_3(r, r_1, r_2, r_3) = \varphi_3(-r, r_1, r_2, r_3)$, as stated earlier. This relation is obvious from the fact that polynomial g_i in Equation (16) has the same form as f_i in Equation (15) if we substitute $y_i' = y_i - d$, $r' = -r$. Under these substitutions, polynomial h_i in Equation (17) remains the same. Therefore, we need only to compute $\varphi_n(R, r_i)$ considering the ideal $J_R = (f_i, h_i)$ for $n \geq 4$ hereafter.

3.3 Computation for $n = 4$

Computation of the ideal $J = (f_1, f_2, f_3, f_4, g_1, g_2, g_3, g_4, h_1, h_2, h_3, h_4)$ gives several relations such as $R^2 - 6Rr + r^2 - d^2$ in Equation (12) and $r_2 r_3 r_4 + r_1 r_2 r_4 - r_1 r_3 r_4 - r_1 r_2 r_3$ in Equation (13) by the polynomials in the Gröbner basis and their factors. We omit the details here, and focus on the computation of $\varphi_4(R, r_i)$.

Letting $J_R = (f_1, f_2, f_3, f_4, h_1, h_2, h_3, h_4)$, we compute its Gröbner basis using the ordering $[x_1, x_2, x_3, x_4, y_1, y_2, y_3, y_4] \succ [R, r_1, r_2, r_3, r_4]$. Then, the basis contains a polynomial of the form $R^2(r_2 - r_4)(r_1 - r_3) \times \varphi_4(R, r_i)$, where

$$\begin{aligned}
\varphi_4(R, r_i) &= ((r_1 r_2 - r_2 r_3 + r_3 r_4 - r_4 r_1)^2 - 16 r_1 r_2 r_3 r_4) R^4 \\
&\quad + 16(r_1 + r_2 + r_3 + r_4) r_1 r_2 r_3 r_4 R^3 \\
&\quad - 8(2 r_1 r_3 + 2 r_2 r_4 + r_1 r_2 + r_2 r_3 + r_3 r_4 + r_4 r_1) r_1 r_2 r_3 r_4 R^2 \\
&\quad + 16 r_1^2 r_2^2 r_3^2 r_4^2 \\
&= 0.
\end{aligned} \tag{18}$$

If we let $\psi_4(r, r_i) = \varphi_4(-r, r_i)$, then quartic equations $\varphi_4(R, r_i) = 0$ and $\psi_4(r, r_i) = 0$ could be equivalent to $\varphi_3(R, r_i) = 0$ and $\psi_3(r, r_i) = 0$ in Equations (8) and (9). We investigate them in the following subsection.

3.4 Discussion on the Formulae for $n = 4$

First, we transform the equations $\varphi_4(R, r_i) = 0$ and $\psi_4(r, r_i) = 0$ into equations with K, κ, and κ_i, substituting $K = 1/R$, $\kappa = 1/r$, and $\kappa_i = 1/r_i$. We then have the following polynomial relations among the curvatures:

$$\begin{aligned}
\Phi_4(K, \kappa_i) &= 16 K^4 - 8 \left(2\kappa_1\kappa_3 + 2\kappa_2\kappa_4 + \kappa_1\kappa_2 + \kappa_2\kappa_3 + \kappa_3\kappa_4 + \kappa_4\kappa_1\right) K^2 \\
&\quad + 16 \left(\kappa_1\kappa_2\kappa_3 + \kappa_1\kappa_2\kappa_4 + \kappa_1\kappa_3\kappa_4 + \kappa_2\kappa_3\kappa_4\right) K \\
&\quad + \left(\kappa_1\kappa_2 - \kappa_2\kappa_3 + \kappa_3\kappa_4 - \kappa_4\kappa_1\right)^2 - 16\kappa_1\kappa_2\kappa_3\kappa_4 \\
&= 0,
\end{aligned} \tag{19}$$

$$\begin{aligned}
\Psi_4(\kappa, \kappa_i) &= 16\kappa^4 - 8 \left(2\kappa_1\kappa_3 + 2\kappa_2\kappa_4 + \kappa_1\kappa_2 + \kappa_2\kappa_3 + \kappa_3\kappa_4 + \kappa_4\kappa_1\right) \kappa^2 \\
&\quad - 16 \left(\kappa_1\kappa_2\kappa_3 + \kappa_1\kappa_2\kappa_4 + \kappa_1\kappa_3\kappa_4 + \kappa_2\kappa_3\kappa_4\right) \kappa \\
&\quad + \left(\kappa_1\kappa_2 - \kappa_2\kappa_3 + \kappa_3\kappa_4 - \kappa_4\kappa_1\right)^2 - 16\kappa_1\kappa_2\kappa_3\kappa_4 \\
&= 0.
\end{aligned} \tag{20}$$

Second, we eliminate, for example, κ_4 in Equations (19) and (20), because we have $\kappa_1 + \kappa_3 = \kappa_2 + \kappa_4$ from Equation (13). If we substitute $\kappa_4 = \kappa_1 + \kappa_3 - \kappa_2$ into $\Phi_4(K, \kappa_i)$ and $\Psi_4(\kappa, \kappa_i)$, we obtain

$$(2K - (\kappa_1 + \kappa_3))^2 (4K^2 + 4(\kappa_1 + \kappa_3)K + \kappa_1^2 + 4\kappa_2^2 + \kappa_3^2 - 2(2\kappa_1\kappa_2 + 2\kappa_2\kappa_3 + \kappa_3\kappa_1)) = 0, \tag{21}$$

$$(2k + (\kappa_1 + \kappa_3))^2 (4k^2 - 4(\kappa_1 + \kappa_3)k + \kappa_1^2 + 4\kappa_2^2 + \kappa_3^2 - 2(2\kappa_1\kappa_2 + 2\kappa_2\kappa_3 + \kappa_3\kappa_1)) = 0. \tag{22}$$

Since $\kappa = -(\kappa_1 + \kappa_3)/2\ (< 0)$ is a spurious solution, the essential parts of these equations are given by

$$4K^2 + 4(\kappa_1 + \kappa_3)K + \kappa_1^2 + 4\kappa_2^2 + \kappa_3^2 - 2(2\kappa_1\kappa_2 + 2\kappa_2\kappa_3 + \kappa_3\kappa_1) = 0, \quad (23)$$

$$4k^2 - 4(\kappa_1 + \kappa_3)k + \kappa_1^2 + 4\kappa_2^2 + \kappa_3^2 - 2(2\kappa_1\kappa_2 + 2\kappa_2\kappa_3 + \kappa_3\kappa_1) = 0, \quad (24)$$

which correspond to Equations (3) and (4) in the case $n = 3$.

Finally, if we solve Equations (23) and (24) with K and κ and adopt suitable solutions, we obtain

$$K = -\frac{\kappa_1 + \kappa_3}{2} + \sqrt{\kappa_1\kappa_2 + \kappa_2\kappa_3 + \kappa_3\kappa_1 - \kappa_2^2}, \quad (25)$$

$$k = \frac{\kappa_1 + \kappa_3}{2} + \sqrt{\kappa_1\kappa_2 + \kappa_2\kappa_3 + \kappa_3\kappa_1 - \kappa_2^2}, \quad (26)$$

where $\kappa - K = \kappa_1 + \kappa_3$ satisfies Equation (13). These results are rather simple but do not seem to have been explicitly introduced before. Hence, we could call Equations (25) and (26) Steiner's formulae in the case $n = 4$, correspondently to Equations (5) and (6) in the case $n = 3$.

Example 1. *We let $r_1 = r_2 = r_3 = 1$, then we have $r_4 = 1$. Equations (23) and (24) give such solutions, if we write them in the radii, as $R = \{1 - \sqrt{2},\ 1 + \sqrt{2}\}$ and $r = \{-1 - \sqrt{2},\ -1 + \sqrt{2}\}$. Therefore, we have the solutions $R = 1 + \sqrt{2}$ and $r = -1 + \sqrt{2}$, which form a congruent Steiner 4-cycle.*

Example 2. *Assume that we are given $r_1 = 3$, $r_2 = 2$, and $r_3 = 1$. Then we have $r_4 = 6/5$ by Equation (13). If we solve Equations (23) and (24) and compute their reciprocals, we have $R = \{-0.6524,\ 5.0161\}$ and $r = \{-5.0161,\ 0.6524\}$. Therefore, we have the solutions $R = 5.0161$ and $r = 0.6524$.*

Remark 1. *The above Example 2 is a problem solved by Zen Hodoji, a 19th century mathematician in Japan [5][12]. After obtaining $r_4 = 1.2$, he solved the following equation:*

$$\{(r_4 - r_2)^2 r_1{}^2 - 4(r_2 + r_4)r_1 r_2 r_4 + 4r_2{}^2 r_4{}^2\} R^2$$
$$+ 4(r_2 + r_4)r_1{}^2 r_2 r_4 R + 4r_1{}^2 r_2{}^2 r_4{}^2 = 0, \quad (27)$$

which gives $R = \{-0.6524,\ 5.0161\}$. That is to say, Equation (27) is equivalent to Equation (23), but it has not yet been clarified how Hodoji derived this equation.

4 Computation by Resultants

4.1 Computation for $n = 5$

For the cases of $n \geq 5$, it seems impossible to compute the Gröbner basis directly, using our present computational environment (Maple 14 in Win64, Xeon

(2.93 GHz)×2, 96 GB RAM). Since Maple 15 and Maple 16 behaved somehow inefficiently with respect to our problems, we describe the computation time obtained by Maple 14 in the following.

Instead of the Gröbner basis, resultant methods are often more useful for computational geometry problems. Analogously with our previous paper [9], we try to eliminate x_i, y_i in the system of Equations (15) and (17) by resultants. Actually, we consider the following system with ten polynomials:

$$
\begin{cases}
f_1 = x_1^2 + y_1^2 - (R - r_1)^2 \\
f_2 = x_2^2 + y_2^2 - (R - r_2)^2 \\
f_3 = x_3^2 + y_3^2 - (R - r_3)^2 \\
f_4 = x_4^2 + y_4^2 - (R - r_4)^2 \\
f_5 = x_5^2 + y_5^2 - (R - r_5)^2 \\
h_1 = (x_2 - x_1)^2 + (y_2 - y_1)^2 - (r_2 - r_1)^2 \\
h_2 = (x_3 - x_2)^2 + (y_3 - y_2)^2 - (r_3 - r_2)^2 \\
h_3 = (x_4 - x_3)^2 + (y_4 - y_3)^2 - (r_4 - r_3)^2 \\
h_4 = (x_5 - x_4)^2 + (y_5 - y_4)^2 - (r_5 - r_4)^2 \\
h_5 = (x_1 - x_5)^2 + (y_1 - y_5)^2 - (r_1 - r_5)^2,
\end{cases}
\tag{28}
$$

whose variables are eliminated as follows:

(i) Without loss of generality, we can fix the center of the first inner circle as $x_1 = 0$ and $y_1 = R - r_1$. If we substitute them to the polynomials f_1, h_1, and h_5, all x_1 and y_1 in the system are eliminated and we obtain $f_1 \to 0$.

(ii) Computing $h_1 - f_2$ yields a linear equation with y_2. Hence, we substitute the solution y_2 to f_2 and h_2, cancel the denominators, and let them be $f_2^{(1)}$ and $h_2^{(1)}$, respectively. In this step, we have $h_1 \to 0$.

(iii) Computing $h_5 - f_5$ yields a linear equation with y_5. Hence, we substitute the solution y_5 to f_5 and h_4, cancel the denominators, and let them be $f_5^{(1)}$ and $h_4^{(1)}$, respectively. In this step, we have $h_5 \to 0$.

(iv) At this point, the following seven polynomials are left as non-zero:

$$f_2^{(1)}(x_2, R, r_i), \ f_3(x_3, y_3, R, r_i), \ f_4(x_4, y_4, R, r_i), \ f_5^{(1)}(x_5, R, r_i),$$

$$h_2^{(1)}(x_2, x_3, y_3, R, r_i), \ h_3(x_3, x_4, y_3, y_4, R, r_i), \ h_4^{(1)}(x_4, x_5, y_4, R, r_i).$$

(v) We compute the resultants to eliminate each variable in order. Factoring the resultants, we pick up only the essential factors:

$$h_2^{(2)}(x_3, y_3, R, r_i) \leftarrow \mathrm{Res}_{x_2}(h_2^{(1)}, f_2^{(1)}), \quad h_2^{(3)}(y_3, R, r_i) \leftarrow \mathrm{Res}_{x_3}(h_2^{(2)}, f_3).$$

(vi) Analogously, we compute

$$h_3^{(1)}(x_4, y_3, y_4, R, r_i) \leftarrow \mathrm{Res}_{x_3}(h_3, f_3), \quad h_3^{(2)}(y_3, y_4, R, r_i) \leftarrow \mathrm{Res}_{x_4}(h_3^{(1)}, f_4).$$

(vii) Analogously, we compute

$$h_4^{(2)}(x_4, y_4, R, r_i) \leftarrow \mathrm{Res}_{x_5}(h_4^{(1)}, f_5^{(1)}), \quad h_4^{(3)}(y_4, R, r_i) \leftarrow \mathrm{Res}_{x_4}(h_4^{(2)}, f_4).$$

(viii) Moreover, we compute

$$h_3^{(3)}(y_4, R, r_i) \leftarrow \operatorname{Res}_{y_3}(h_2^{(3)}, h_3^{(2)}), \quad \varphi_5(R, r_i) \leftarrow \operatorname{Res}_{y_4}(h_3^{(3)}, h_4^{(3)}),$$

which finally gives the relation among $R, r_1, \ldots, r_5$:

$$\varphi_5(R, r_i) = a_{24}(r_i)R^{24} + \cdots + a_1(r_i)R + a_0(r_i) = 0 \quad (a_j(r_i) \in \mathbf{Q}[r_1, \ldots, r_5]).$$
$$(29)$$

We applied these steps to the cases $n = 3$ and 4, and obtained the same polynomials $\varphi_3(R, r_i)$ and $\varphi_4(R, r_i)$ as were derived from the Gröbner basis computation. Therefore, we believe that the above polynomial $\varphi_5(R, r_i)$ is surely the defining polynomial of R in the case $n = 5$. The CPU time was about 17 seconds for the whole process to compute $\varphi_5(R, r_i)$ and to confirm its irreducibility.

As discussed above, Equation (29) can be transformed into the expression by curvatures as follows:

$$\Phi_5(K, \kappa_i) = \tilde{a}_{24}(\kappa_i)K^{24} + \cdots + \tilde{a}_1(\kappa_i)K + \tilde{a}_0(\kappa_i) = 0 \quad (\tilde{a}_j(\kappa_i) \in \mathbf{Q}[\kappa_1, \ldots, \kappa_5]).$$
$$(30)$$

However, it is not yet possible to analyze $\Phi_5(K, \kappa_i)$ further, because we do not have any relations among $\kappa_1, \ldots, \kappa_5$ analogously to Equations (7) and (13).

Example 3. *If we let $r_1 = r_2 = r_3 = r_4 = r_5 = 1$, then the equation $\varphi_5(R, r_i) = 0$ is factored as follows, and the solutions are divided into four cases:*

$$(R - 1)^{10}\left(R^2 - 2R - \frac{1}{3}\right)^5\left(R^4 - 4R^3 + 2R^2 + 4R + \frac{1}{5}\right) = 0. \quad (31)$$

(i) $R = 1$, $r = -1$.
All the circles with radius 1 coincide at one center.

(ii) $R = 1 + 2/\sqrt{3}$, $r = -1 + 2/\sqrt{3}$.
Five circles are degenerated into a congruent Steiner 3-cycle; that is, three circles are located so that their centers form a regular triangle.

(iii) $R = 1 + \sqrt{2 + 2/\sqrt{5}}$, $r = -1 + \sqrt{2 + 2/\sqrt{5}}$.
Five circles form a truly congruent Steiner 5-cycle; that is, five circles are located so that their centers form a regular pentagon.

(iv) $R = 1 + \sqrt{2 - 2/\sqrt{5}}$, $r = -1 + \sqrt{2 - 2/\sqrt{5}}$.
Five circles are located so that their centers form a regular pentagram, which could be called a congruent Steiner 5-cycle with 2 revolutions. In general, a Steiner n-cycle with m revolutions satisfies, extended from Equation (10),

$$(R - r)^2 - 4Rr \tan^2 \frac{m\pi}{n} = d^2, \quad (32)$$

and this is the case where $n = 5$ and $m = 2$.

4.2   Computation for $n = 6$

Since Equations (15) and (17) have a systematic structure, we can also prospect the order of elimination of variables for the case $n = 6$. Actually, we succeeded in computing the polynomial $\varphi_6(R, r_i)$, which has degree 48 in R:

$$\varphi_6(R, r_i) = b_{48}(r_i)R^{48} + \cdots + b_1(r_i)R + b_0(r_i) = 0 \qquad (b_j(r_i) \in \mathbf{Q}[r_1, \ldots, r_6]).$$
(33)

The CPU time was 7 hours and 40 minutes, but the elapsed time was about 1 hour and 6 minutes because of multithreaded computation.

Next, we found that this polynomial $\varphi_6(R, r_i)$ is factored as

$$\varphi_6(R, r_i) = \varphi_6^{(1)}(R, r_i) \cdot \varphi_6^{(2)}(R, r_i),$$
(34)

where the degrees in R of $\varphi_6^{(1)}(R, r_i)$ and $\varphi_6^{(2)}(R, r_i)$ are 16 and 32, respectively. This step took 34 hours and 27 minutes of CPU time (22 hours and 35 minutes of elapsed time) and 77 GB of RAM.

In order to compute the polynomial equation in the curvatures, we substituted $K = 1/R$, $\kappa = 1/r$, and $\kappa_i = 1/r_i$ to $\varphi_6^{(1)}(R, r_i)$ and $\varphi_6^{(2)}(R, r_i)$. Moreover, two of the six κ_i's are supposed to be eliminated; for example, by substituting $\kappa_6 = \kappa_1 + \kappa_4 - \kappa_3$ and $\kappa_5 = \kappa_1 + \kappa_4 - \kappa_2$ from Equation (13). After simplifying the two polynomials in K and the κ_i's, we obtained the following equations in a factored form:

$$\begin{cases} (2K - (\kappa_1 + \kappa_4))^4 \cdot \Phi_6^{(1)}(K, \kappa_i) = 0 \\ \Phi_6^{(2)}(K, \kappa_i) = 0, \end{cases}$$
(35)

where the degrees in K of $\Phi_6^{(1)}(K, \kappa_i)$ and $\Phi_6^{(2)}(K, \kappa_i)$ are 12 and 32, respectively.

Example 4. *If we let $\kappa_1 = \kappa_2 = \kappa_3 = \kappa_4 = 1$, then equation $\Phi_6^{(1)}(K, \kappa_i) = 0$ is factored as follows:*

$$(3K - 1)(K + 1)(2K - 1)^4(K - 1)^6 = 0.$$
(36)

The case $K = 1/3$, that is, $R = 3$, means a congruent Steiner 6-cycle, and other roots $R = \pm 1$ and $R = 2$ correspond to degenerated cases.

Similarly, equation $\Phi_6^{(2)}(K, \kappa_i) = 0$ is factored if $\kappa_i = 1$,

$$(K^2 + 6K - 3)(K^2 + 2K - 1)^6(K - 1)^{18} = 0.$$
(37)

The solutions other than $K = 1$ are correspondent to the following cases:

(i) *$R = 1 + 2/\sqrt{3}$, $r = -1 + 2/\sqrt{3}$ (from $K = -3 \pm 2\sqrt{3}$).*
   Three circles are located so that their centers form a regular triangle (a congruent Steiner 3-cycle).
(ii) *$R = 1 + \sqrt{2}$, $r = 1 - \sqrt{2}$ (from $K = -1 \pm \sqrt{2}$).*
   Four circles are located so that their centers form a regular square (a congruent Steiner 4-cycle).

Table 1. Defining polynomial $\varphi_n(R, r_i)$ of radius R for a Steiner n-cycle

n	Degree in R	No. of terms
3	2	10
4	4	20
5	24	30,645
6	16	271,828
	32	310,935

5   Concluding Remarks

In this study, we tried to find some new formulae for a Steiner n-cycle using the Gröbner basis and resultant method for the equations of coordinates of circles.

For the case $n = 3$, the Descartes circle theorem is deduced by Gröbner basis computation, which means that another proof without inversion for the theorem is given.

For the case $n = 4$, we have obtained two quartic equations, Equations (19) and (20), which can be regarded as the extension of the Descartes circle theorem to $n = 4$. However, using the relation $\kappa_1 + \kappa_3 = \kappa_2 + \kappa_4$, solutions K and κ are simply expressed by Equations (25) and (26). It seems rather peculiar if these results have not been pointed out before.

For the cases $n = 5$ and $n = 6$, it was impossible to compute the Gröbner basis. Instead, using the resultant method, we succeeded in computing the final relations $\varphi_5(R, r_i) = 0$ and $\varphi_6^{(1)}(R, r_i) \cdot \varphi_6^{(2)}(R, r_i) = 0$. These formulae are also expressed by the polynomials in the curvatures, as $\Phi_5(K, \kappa_i) = 0$, $\Phi_6^{(1)}(K, \kappa_i) = 0$, and $\Phi_6^{(2)}(K, \kappa_i) = 0$.

We summarize the shapes of $\varphi_n(R, r_i)$ for $n = 3, 4, 5$, and 6 in Table 1. To the best of our knowledge, there exist no other reports in which these formulae are explicitly given. However, it has not yet been sufficiently elucidated whether Equations (30) and (35) have geometrical significance other than for congruent Steiner n-cycle cases. Therefore, they should be clarified in future work for generic cases, together with the following problems whose studies are still ongoing.

(i) Apparently, it might be difficult to compute the case $n = 7$, because the size of the polynomial $\varphi_6(R, r_i)$ reaches the limit of the present computational resources (Xeon (2.93 GHz)×2, 96 GB RAM). Using other computer algebra systems could improve the computation time, but the space complexity gives rise to the essential difficulty.

(ii) We have some evidence that the degree in R of $\varphi_7(R, r_i)$ should be 160, which is obtained under such substitution by random prime numbers as $r_i = p_i$. Hence, we believe that the sequence of degrees $2, 4, 24, 48, 160, \ldots$ should have some combinatorial meaning.

(iii) Another historical Japanese mathematician, Naonobu Ajima, derived the following recurrence relation in his book *Renjutsu-Henkan* in 1784 [8]:

$$\frac{1}{r_n} = \frac{1}{r_{n-1}} \left(\frac{16Rr}{(R+r)^2 - d^2} - 2 \right) + \frac{8(R-r)}{(R+r)^2 - d^2} - \frac{1}{r_{n-2}}. \tag{38}$$

Using this relation, he computed the radii $r_3, r_4, \ldots$ in order for given R, r, r_1, and r_2 in a Steiner n-cycle. Finally, he pointed out that, if the chain of n circles is closed, we should have

$$\frac{4Rr}{(R+r)^2 - d^2} = \cos^2 \frac{\pi}{n}, \tag{39}$$

which is equivalent to Equation (10) given by Jakob Steiner (1796-1863). Therefore, we should consider whether a general term expressed by Equation (38) could be helpful for describing the geometric meaning.

(iv) There exist some related works that discuss extension to the conic case [7] or 3D case [11]. Both papers apply numerical computations when symbolic computation is impossible. We should try to apply our approach to those problems, but it still seems very difficult to compute them fully symbolically.

References

1. Chou, S.C.: Mechanical Geometry Theorem Proving, D. Reidel, Dordrecht (1988)
2. Coxeter, H.S.M.: The Problem of Apollonius. American Mathematical Monthly 75(1), 5–15 (1968)
3. Coxeter, H.S.M., Greitzer, S.L.: Geometry Revisited. MAA, Washington, D.C. (1967)
4. Forder, H.G.: Geometry. Hutchinson, London (1960)
5. Fukagawa, H., Rothman, T.: Sacred Mathematics: Japanese Temple Geometry. Princeton (2008)
6. Lagarias, J.C., Mallows, C.L., Wilks, A.R.: Beyond the Descartes Circle Theorem. American Mathematical Monthly 109(4), 338–361 (2002)
7. Lewis, R.H., Bridgett, S.: Conic Tangency Equations and Apollonius Problems in Biochemistry and Pharmacology. Mathematics and Computers in Simulation 61(2), 101–114 (2003)
8. Matsuoka, M.: Explanations on Naonobu Ajima's Complete Works. Fuji Jr. College, Tokyo (1966)
9. Moritsugu, S.: Computing Explicit Formulae for the Radius of Cyclic Hexagons and Heptagons. Bulletin of Japan Soc. Symbolic and Algebraic Computation 18(1), 3–9 (2011)
10. Pedoe, D.: On a Theorem in Geometry. American Mathematical Monthly 74(6), 627–640 (1967)
11. Roanes-Macías, E., Roanes-Lozano, E.: 3D Extension of Steiner Chains Problem. Mathematical and Computer Modelling 45(1,2), 137–148 (2007)
12. Tanaka, M., Kobayashi, M., Tanaka, M., Ohtani, T.: On Hodohji's Sanhenpou Method: Inversion Theorem Formulated by Traveling Mathematician in Old Japan. J. of Education for History of Technology 7(2), 28–33 (2006) (in Japanese)
13. Wilker, J.B.: Four Proofs of a Generalization of the Descartes Circle Theorem. American Mathematical Monthly 76(3), 278–282 (1969)

Equation Systems
with Free-Coordinates Determinants

Pascal Mathis and Pascal Schreck

LSIIT, UMR CNRS 7005
Université de Strasbourg, France
{mathis,schreck}@unistra.fr

Abstract. In geometric constraint solving, it is usual to consider Cayley-Menger determinants in particular in robotics and molecular chemistry, but also in CAD. The idea is to regard distances as coordinates and to build systems where the unknowns are distances between points. In some cases, this allows to drastically reduce the size of the system to solve. On the negative part, it is difficult to know in advance if the yielded systems will be small and then to build these systems. In this paper, we describe two algorithms which allow to generate such systems with a minimum number of equations according to a chosen reference with 3 or 4 fixed points. We can then compute the smaller systems by enumeration of references. We also discuss what are the criteria so that such system can be efficiently solved by homotopy.

1   Introduction

The main goal of distance geometry consists in specifying sets of points by the mean of known pairwise distances. This issue is widely studied in domains where distances are primitive notions such as molecular modeling in chemistry or biology [6], robotics [11], cartography with town and country planning [1] or even in CAD [5] and mathematics [14]. Distance geometry allows to fundamentally incorporate in the formulation of the problem the invariance by the group of direct isometries. In the particular field of constraint solving, considering distance geometry —when it is possible— can drastically reduce the size of the equation system to solve.

One of the classical tools used in distance geometry is based on Cayley-Menger determinants. Some authors studied extensions in order to take into account other geometric objects and relations such points and distances. For instance, in [5,14], hyperplanes and hyperspheres are considered together with constraints of incidence, tangency and angle. These results are encouraging steps toward the use of distance geometry in CAD, even if one cannot yet manage lines in 3D within this framework.

When considering geometric constraint solving in CAD, methods based on Cayley-Menger determinants have to be mixed with other ingredients. For instance, trilateration (intersection of hyper-spheres) has to be used to retrieve the coordinates of points, or decomposition of large constraint systems into smaller

T. Ida and J. Fleuriot (Eds.): ADG 2012, LNAI 7993, pp. 59–70, 2013.
© Springer-Verlag Berlin Heidelberg 2013

rigid subsystems must be done in order to manage the complexity [4]. But, one must bear in mind that the systems of algebraic equations yielded by using Cayley-Menger have to be effectively solved. Thus, it is important to have systems with the smallest possible size and/or the smallest possible total degree.

In this paper, we propose two algorithms that transform a geometric constraint system involving points and hyperplanes with distance and angle constraints, into an equivalent system of Cayley-Menger determinants. These algorithms are interesting for small indecomposable constraint systems since they produce one of the smallest systems of Cayley-Menger determinant obtainable by fixing a reference of four or three points. We also describe a brute force algorithm and a greedy heuristic to find the better reference(s) for a given problem. We focus on 3D problems, but it could be used in 2D for indecomposable problems. We present some examples using this method with the standard homotopy method for numerical solving.

The rest of the paper is organized as follows. Section 2 introduces Cayley-Menger determinants and their extensions. Section 3 describes our algorithms. Section 4 presents some experiments that we made, and the results that we obtained. We conclude in section 5 by giving some tracks for future works.

2   Cayley-Menger Determinants

Given n points $\{p_1, \ldots, p_n\}$ in the Euclidean space of dimension d, the Cayley-Menger determinant (CMD for short) of these points is:

$$D(p_1, \ldots, p_n) = \begin{vmatrix} 0 & 1 & 1 & \ldots & 1 \\ 1 & 0 & r_{1,2} & \ldots & r_{1,n} \\ 1 & r_{2,1} & 0 & \ldots & r_{2,n} \\ \vdots & \vdots & \vdots & \ddots & \vdots \\ 1 & r_{n,1} & r_{n,2} & \ldots & 0 \end{vmatrix}$$

where $r_{i,j}$ is the square distance between p_i and p_j.

In dimension d, a set of $n \geq d+2$ points specified by a Cayley-Menger determinant is embeddable in $\mathbb{R}^d$ if $D(p_1, \ldots, p_n) = 0$. In particular, in 3D, for 5 or 6 distinct points we have : $D(p_1, p_2, p_3, p_4, p_5) = 0$ and $D(p_1, p_2, p_3, p_4, p_5, p_6) = 0$.

Consider the 2D example of Figure 1, CMD of points $p1, \ldots, p4$ is :

$$D(p1, p2, p3, p4) = \begin{vmatrix} 0 & 1 & 1 & 1 & 1 \\ 1 & 0 & 36 & 74 & 20 \\ 1 & 36 & 0 & 26 & r_{2,4} \\ 1 & 74 & 26 & 0 & 82 \\ 1 & 20 & r_{2,4} & 82 & 0 \end{vmatrix} = -148(r_{2,4})^2 + 14560(r_{2,4}) - 217600$$

This determinant contains one unknown $r_{2,4}$ because this is the only missing distance among all possible distances of these four points. Since the points are coplanar, the relation $D(p1, p2, p3, p4) = 0$ must hold. Solving the quadratic equation gives $r_{2,4} = 80$ or $r_{2,4} = 680/37$. So distance $p2p4$ must be $\sqrt{80}$ or

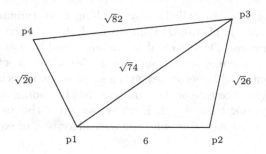

Fig. 1. Simple distance configuration in 2D

$\sqrt{\frac{680}{37}}$ in order that the four points lie on a plane. Other numerical values for distance $p2p4$ correspond to non flat tetrahedron configuration.

In addition, for $n = d + 3$, Sippl and Scheraga [7] showed that Cayley-Menger determinant equation for points and distances $D(p_1, \ldots, p_{n-1}, p_n) = 0$ can be substituted by $D^*(p_1, \ldots, p_{n-1}, p_n) = 0$ where:

$$D^*(p_1, \ldots, p_{n-1}, p_n) = \begin{vmatrix} 0 & 1 & 1 & \ldots & 1 \\ 1 & 0 & r_{1,2} & \ldots & r_{1,n-1} \\ 1 & r_{2,1} & 0 & \ldots & r_{2,n-1} \\ \vdots & \vdots & \vdots & \ddots & \vdots \\ 1 & r_{n-2,1} & r_{n-2,2} & \vdots & r_{n-2,n-1} \\ 1 & r_{1,n} & r_{2,n} & \ldots & r_{n-1,n} \end{vmatrix}$$

The indexing notation D^* is used by most authors to indicate Sippl optimization. This makes reading the two types (5 points and 6 points) of 6x6 determinants easier in systems. The Sippl optimization that removes one line and one column is made thanks to Jacobi's theorem with the fact that minors involving $d + 2$ objects are equal to 0. It will be useful to minimize the degree of the equations.

Cayley-Menger determinants were extended by [13] to hyperplanes and hyperspheres. In this paper, we use definitions coming from [13] but we restrict ourselves to points and hyperplanes. For a given set of n objects p_i, the matrix M is defined as follows:

$$M = \begin{pmatrix} 0 & \delta \\ \delta^t & G \end{pmatrix}$$

with $\delta = (\delta_1, \ldots, \delta_n)$ and $\delta_i = 1$ if p_i is a point and $\delta_i = 0$ if it is a hyperplane. Noting $g_{i,j}$ the i-th row and j-th column element of G, we have :

- $g_{i,j}$ is the square distance between p_i and p_j if they are both points
- $g_{i,j}$ is the signed distance of p_i and p_j if one is a point and the other a hyperplane (so 0 for incidence relation)
- $g_{i,j}$ equals $-\frac{1}{2}cos(p_i, p_j)$ if they are both hyperplanes.

The determinant of M is also called a Cayley-Menger determinant and the previous notation $D(p_1, \ldots, p_n)$ is used. In dimension d, the property $D(p_1, \ldots, p_n) = 0$ if $n \geq d + 2$ remains. Thus, such determinants lead to equations where unknowns correspond to unknown distances or angles. Given a set O of geometric objects among points and hyperplanes, and a set of constraints among distances, incidences and angles, the rule of the game consists in finding subsets of O, each of size equal or greater than $d + 2$. Each subset gives rise to the nullity of a determinant, that is an equation, and the overall system of equations must be well-formed.

3   Two Algorithms for Setting CM-Systems

We call *CM-system* related to system S any equational system where every equation is a CMD. One of the crucial questions when dealing with CM-system is: "how can I compute a reasonable CM-system equivalent to S?".

We describe here two algorithms which have good qualities for small constraint systems. Since each considered constraint involves two objects (point or hyperplane), system S can be seen as a constraint graph G. Its vertices correspond to the points or planes and its edges to given distances.

In both algorithms a subset of 3 or 4 points must be given. Such a subset is called a *reference*. All edges where end-points are not in current reference are said *external edges*. Consider example of figure 2a that is the graph for the disulfide bond such given in [6]. For reference $\{p1, p2, p3\}$, there are 8 external edges which are shown in dotted lines in 2b.

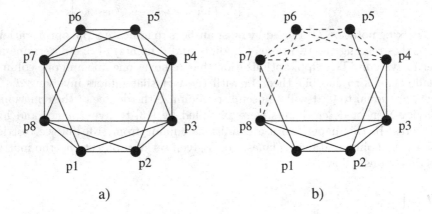

Fig. 2. a) Disulfide bond , b) external edges in dotted lines for reference $\{p1, p2, p3\}$

Each algorithm inputs a well-constrained constraint system S (or equivalently its graph) and a reference R. In the 3D case, the first algorithm considers references with 4 points, and the second one uses references of 3 points.

3.1   A 3D Algorithm with $|R| = 4$

Without loss of generality, we note the points $\{p_1, p_2, \ldots p_n\}$, with $R = \{p_1, p_2, p_3, p_4\}$. The first algorithm is described by the following pseudo-code:

Input: G and $R = \{p_1, \ p_2, \ p_3, \ p_4\}$
Output: SCM
SCM = $\emptyset$
1- for each point P in $\{p_5, \ \ldots \ p_n\}$, add $D(R, P)$ to SCM
2- for each external edge $(P, \ Q)$ of G, add $D(R, P, Q)$ to SCM

Let consider example of Figure 2. With reference $R = \{p_1, p_2, p_3, p_8\}$, 4 points $\{p_4, p_5, p_6, p_7\}$ are not in R, each point gives rise to a CMD. There are 6 external edges, each of them is in a CMD of 6 points. This leads to a CM-system of 10 equations:

$$\begin{cases} D(R, p_4) & = 0 \\ D(R, p_5) & = 0 \\ D(R, p_6) & = 0 \\ D(R, p_7) & = 0 \\ D^*(R, p_4, p_5) = 0 \\ D^*(R, p_4, p_6) = 0 \\ D^*(R, p_4, p_7) = 0 \\ D^*(R, p_5, p_6) = 0 \\ D^*(R, p_5, p_7) = 0 \\ D^*(R, p_6, p_7) = 0 \end{cases}$$

The first equation, $D(R, p_4) = 0$, involves all possible pairwise distances between points p_1, p_2, p_3, p_8, p_4. Among these 10 distances, only two of them, $p_1 p_4$ and $p_4 p_8$ are not known, they do not correspond to edges in graph of Figure 2. So the first equation has two unknowns. And so on with others CMD.

Once solved, the solutions of this CM-system allow to compute the coordinates for points of S. First, the tetrahedron reference is constructed with arbitrarily chosen coordinates that fulfilled the distance constraints. Then, others coordinates of points are easily computed since for each point p among $\{p_4, p_5, p_6, p_7\}$, distances between p and points of reference are known. This way, all the points are expressed within this reference and trilateration can be used to compute the coordinates of all the points. Notice that all the distances of S occur in SCM.

We say that such a system is CM-equivalent to S if it is well-constrained and it allows to retrieve all the solutions of S.

First, we prove that SCM is well-constrained when S is so. It is worth to recall that well-constrained means "well-constrained modulo the direct isometries" for S while it means "having a finite number of solutions for the unknown distances" for SCM.

We first show that SCM is structurally well-constrained, that is: (a) it contains as many equations as unknowns; and (b) each of its subsystems contains more unknowns than equations.

For point (a), let us note t the total number of equations of SCM, n the number of points of S and k the number of external edges. It is easy to see that $t = (n-4) + k$. On the other hand, we have $4(n-4)$ distances involved for each point P in the step 1 (in addition to the 6 distances involved in completing R to a clique in G), and k additional distances involved in the second step. Summing up the distances, we have $4(n-4) + k + 6$ distances involved in SCM, and since S is well-constrained, it contains $3n-6$ constraints. Then we have, $4(n-4) + k + 6 - 3n + 6 = n - 4 + k$ unknowns and the same number of equations.

For point (b), we consider a sub-system of SCM with m equations involving c constraints and y points of S. Let us note x the number of unknowns (that is the distances which are not constrained by S), and $m = m_1 + m_2$ where m_1 is the number of equations of type 1, coming from step 1, and m_2 is the number of equations of type 2, coming from step 2 of the previous algorithm. We have to prove that $m \leq x$. We have first the equality

$$x = 4(y-4) + m_2 + 6 - c$$

indeed, we have $y-4$ points out of R, for each of these points there is 4 distances linking it to R; in addition, we have m_2 distances corresponding to the external edges from the m_2 equations of type 2. We also have 6 distances to make R a clique and we have to subtract the c distances corresponding to some constraints in S. But, since S is well-constrained, we have also $c \leq 3y - 6$ Thus, we obtain:

$$x \geq y - 4 + m_2$$

It is easy to see that $y - 4 \geq m_1$ ($y - 4 = m_1$ when considering the whole system). We can then conclude that $x \geq m$

Actually, we can directly prove that SCM is CM-equivalent to S. Indeed, SCM is not over-constrained, since every solution of S give a solution of SCM. In turn, because of the step 1 of our algorithm, every solution of SCM gives a set of points which fulfills all the constraints of S since all the constraints of S are taken into account thanks to steps 1 and 2.

3.2   A 3D Algorithm with $|R| = 3$

The second algorithm is even simpler. Three points are chosen for R, and we note $R = \{p_1, p_2, p_3\}$:

Input: G and R
Output: SCM
SCM = $\emptyset$
1- for each external edge (P, Q), add $D(R, P, Q)$ to SCM

In previous example, Figure 2b shows that with $R = \{p_1, p_2, p_3\}$ this algorithm gives a CM-system with 8 equations for solving S:

$$\begin{cases} D(R, p_4, p_5) = 0 \\ D(R, p_4, p_6) = 0 \\ D(R, p_4, p_7) = 0 \\ D(R, p_5, p_6) = 0 \\ D(R, p_5, p_7) = 0 \\ D(R, p_6, p_7) = 0 \\ D(R, p_6, p_8) = 0 \\ D(R, p_7, p_8) = 0 \end{cases}$$

We also note k the number of external edges. SCM contains then k equations. For each point which is not in R, either it is not involved in an external edge and it is connected with three known distances to R, so it can be forgotten, either it is an extremity of an external edge and step 1 add 3 distances between it and R. Moreover all the external edges are considered during this step. So, all the constraints of S are taken into account. Following the reasoning of previous subsection, it is easy to see that SCM contains k equations and $k + 3(n - 3) + 3 - 3n + 6 = k$ unknowns.

One can easily prove that SCM is structurally well-constrained. It is also easy to see that SCM is CM-equivalent to S.

Several questions arise then, among them:

– is algorithm 2 better than algorithm 1?
– do these algorithms yield the better systems CM-equivalent to S?
– how to choose reference R in order to minimize k?
– what are the criteria that can be used to define the "quality" of such a CM-system?

We give a short answer to question 2: no. Consider the example of the double Stewart's platform, the constraint graph is given on Fig. 3 (all the edges correspond to distance constraints). When choosing a single reference, the best choice for algorithm 1 leads to a CM-system with 9 equations related to reference $\{p_1, p_5, p_6, p_7\}$, for instance. For algorithm 2, the better CM-system uses reference $\{p_4, p_5, p_6\}$ and contains 6 equations. By changing the reference during the construction of a CM-system, we can obtain a system with only 4 equations:

$$\begin{cases} D(p_2, p_3, p_4, p_6, p_1) = 0 \\ D(p_2, p_3, p_4, p_6, p_5) = 0 \\ D(p_4, p_6, p_8, p_9, p_5) = 0 \\ D(p_4, p_6, p_8, p_9, p_7) = 0 \end{cases}$$

We discuss the other questions in the next subsection.

3.3   A *Better* System?

Better Than the Classical Cartesian Approach? Size is the first quality criterion one can think of for a constraint system. First, one can wonder if this approach is better than the Cartesian one. With $|R| = 3$, the system is smaller

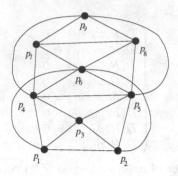

Fig. 3. Double Stewart 's platform

if $k < 3n - 6$ which always occurs (k is the number of external edges which is always less than the overall number of geometric constraints). With $|R| = 4$, it is smaller if $n - 4 + k < 3n - 6$, this implies $k < 2n - 2$ which is often the case. These considerations have to be moderated by the fact that clever methods of decomposition can be used in the classical approaches for geometric constraint solving (see for instance [2,12,10]).

To avoid any ambiguity, in the following, k_3 is k for a 3 points reference and k_4 for a 4 points reference.

A 3 or a 4 Points Reference? Next, one can ask whether it is better to choose either a 3 points or a 4 points reference. Given a 3 points reference R_3, if a 4th vertex p is added, does the number of unknowns could decrease ? With $R_4 = R_3 \cup \{p\}$, it is obvious that $k_3 > k_4$ and that the difference $k_3 - k_4$ represents the degree of vertex p in the graph i.e. the number of incident edges. The number of equations is smaller if $n - 4 + k_4 < k_3$, that implies $n - 4 < k_3 - k_4$. So, if there exist a vertex the degree of which is greater or equal to $n - 4$ then the number of equations can be reduced with $R \cup p$.

Choosing the Best Reference. In short, the number of equations is in $O(k)$. To get the smallest value for k, one has to find out a reference whose vertices cover a maximum set of edges. This is strongly related to the maximum coverage problem which is NP-hard. But for constraint systems involving a small amount of points, a brute force algorithm can be used in practice.

For larger problems, a classical greedy algorithm provides a satisfactory outcome and for most of our examples it gives the best result. This algorithm consists in iteratively choosing one of the most connected vertexes and in removing it from the graph. For instance, with example of disulfide bond and $R = \{p2, p3, p6, p7\}$, the size of the system is 6. With $R = \{p3, p7, p8\}$, its size is 5.

CM-System Degree and Numerical Solving. Once a CM-system has been built, it has to be solved. Here we consider numerical resolution of such systems. CM-systems usually have many solutions, most of which are non-reals (about two thirds of the solutions in our experiments). Indeed, many numerical values found are negative whereas they often represent squared distances. A numerical method like Newton-Raphson has a complexity mostly related to the size of the system. But this method only provides one solution. Since we have algebraic systems of equations, we can use classical homotopy methods in order to get all the solutions.

With homotopy, the criterion is the degree of the system (Bezout bound or BKK bound). Indeed, the total degree of the system determines the number of paths which have to be followed by homotopy. In a polynomial with variables $x_1, \ldots, x_n$ in the form $\sum_i a_i.x_1^{d_1} \ldots x_n^{d_n}$, the degree is the highest value $d_1 + \ldots + d_n$ for a monomial. For a set of m polynomials of degrees $dp_1, dp_2, \ldots dp_m$, the degree of the system is the product $dp_1.dp_2.\ldots.dp_m$. This degree is also called the Bezout bound and is an upper bound of the number of reals solutions. We can notice that in CM-systems with 5 or 6 objects, the degree of each polynomial is 2, 3 or 4. BKK bound is the mixed volume of Newton polytopes formed by polynomials [3]. It is a finner bound than Bezout one, especially when system is sparse.

Here, we have considered Bezout bound. Homotopy complexity is mostly related to the degree of the system but is also linked to the number of equations since in each path the system must be evaluated many times. Finding out the best system is then an optimization problem. It amounts to getting a system with a minimum degree with the least equations.

So far, we performed experiments on problems with less than 30 objects. We use a brute force algorithm which selects the interesting systems with small Bezout bound or with the smaller size. The next section, presents some examples that we have solved with our methods.

4   Examples

As the degree of systems grows exponentially with its size, it is only possible to use homotopy with small systems. This is another reason to have powerful decomposition methods. But, in 3D, indecomposable problems can occur (even if it is quite unusual with CAD problems).

We made some experiments with a kind of brute-force search: on 3D problems, all possible systems with 4 points reference are considered first, then, the same is done with all possible 3 points references. These enumerations are done in $O(n^4)$ with n the number of points. Finally, the best solution according to the degree of the system is selected. Numerical solving is performed with HOM4PS [9]. This free software implements both the classical homotopy continuation method that is based on Bezout bound and the polyhedral homotopy continuation that uses the BKK bound. Packages as Bertini and PHCpack were tested but HOM4PS was more effective and robust. It is faster and always provides the same numerical solutions for a same system.

Table 1 shows results for some few examples. In these examples, numerical values for distances were arbitrarily chosen in small values (distances are between

1 and 10). In this table, r is the reference chosen, eq the number of equations, sol the number of solutions, BB the Bezout bound of the system, $t\ ch$ the solving time in seconds with classical homotopy, BKK the BKK bound and $t\ ph$ the solving time with polyhedral homotopy.

The first example is the common Stewart platform consisting in two triangles lying in two different planes and connecting by six distances. In figure 4a, bold edges are external edges for reference p_2, p_3, p_5. With this very little example, the algorithm provides the same equation systems whatever the user choices : either 4 points or 3 points references, either size criterion or degree criterion.

The example of figure 2 is dealt in [6] (in 825 sec. with a Bernstein solver) with the 10 equations system referenced in the table. The system is "handmade". With a 4 points reference, our brute-force algorithm suggests a system of 6 equations. Nevertheless, the degree of this 6 equations system is higher than the 10 equations one and it takes more time to yield the solutions. With a 3 points reference, the algorithm offers a system having 5 equations. This latter has a higher degree than the 10 equations systems but homotopy complexity also depends on the size of the system. In this case, the 5 equations system takes less time.

Another example comes from [8] and is given in figure 4b. This paper presents a specific parameterization that results in 3 equations. The trick is to decompose the problem into straightforward solved rigid bodies (here the tetrahedrons). The 3 equations are then a formulation of the distances between the top vertex of each tetrahedron and a rear vertex of an neighboring tetrahedron. This formulation takes also into account the shared point between the rigid bodies. Without decomposition, 6 equations are enough to solve it within less than 1 second. The reference chosen by the algorithm is the 3 points shared by the tetrahedrons.

The last example of figure 4c involves 10 points and 24 distances, the dotted lines in the figure represent the edges at the rear.

Table 1. experiments carried out in PC with Intel 2.66 Ghz CPU. Times are given in seconds.

Study case	r	eq	sol	BB	t ch	BKK	t ph
Stewart platform	p_2, p_3, p_5	2	6	16	0.004	8	0.004
Disulfide	p_1, p_2, p_3, p_8	10	18	1024	6.47	64	0.62
Disulfide	p_2, p_3, p_6, p_7	6	18	4096	252.05	496	5.31
Disulfide	p_3, p_7, p_8	5	40	512	3.75	128	0.34
3 tetrahedrons	p_1, p_2, p_3	6	14	512	0.7	128	0.1
4 branches star	p_8, p_9, p_{10}	6	66	4096	15.48	384	1.25

The icosahedron problem [14] is part of the folklore of distance geometry. It consists in 12 points and 30 distances. In the underlying graph, each vertex has degree 5 since each point is linked by 5 distances constraints to 5 neighboring points. With a 3 points reference, the size of the system varies from 15 to 18 equations. As in none configuration of reference there is a vertex with a degree greater or equal to 9, none of 4 points reference can decrease the size of the

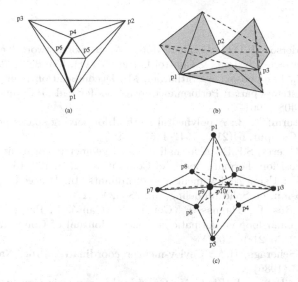

Fig. 4. Few 3D examples

system. Actually, with 4 points reference, the size is between 19 and 23 equations. All these systems have degrees between 2^{30} and 2^{38} that is too much to consider homotopy. This problem cannot be decomposed with common methods used in CAD and such a system must be dealt with other numerical techniques.

5 Conclusion

In this paper, we made a short study on the setting of systems of equations from the Cayley-Menger determinants. The desired quality of the systems depends on the numerical method used to solve it. Here we have considered two criteria: both the size of the system and its degree.

According to our tests, our greedy algorithm produces a good approximation of the smallest CM-system corresponding to a constraint system. For a resolution by homotopy, the degree of the system has a greater influence on performance than the size that it was also involved. Also, we use an algorithm that produces systems both small and with low degree. For now, only a brute-force algorithm has been implemented. We plan to address this optimization problem with conventional methods such as simulated annealing.

Furthermore, in our algorithms, the systems are selected according to Bezout bound but we intend to study a search that results in a system with the smaller BKK bound. Finally, when the number of objects increases the degree becomes prohibitive. However, most problems in CAD can be decomposed into subsystems [4,10]. This decomposition requires the integration of new mechanisms, especially with the choice of the reference.

References

1. Cao, M., Anderson, B.D.O., Stephen Morse, A.: Sensor network localization with imprecise distances. Systems & Control Letters 55(11), 887–893 (2006)
2. Hoffmann, C., Lomonosov, A., Sitharam, M.: Decomposition plans for geometric constraint systems, part i: Performance measures for cad. J. Symbolic Computation 31, 367–408 (2001)
3. Huber, B., Sturmfels, B.: A polyhedral method for solving sparse polynomial systems. Math. Comput. 64(212), 1541–1555 (1995)
4. Mathis, P., Thierry, S.E.B.: A formalization of geometric constraint systems and their decomposition. Formal Aspects of Computing 22(2), 129–151 (2010)
5. Michelucci, D.: Using cayley menger determinants. In: Proceedings of the 2004 ACM Symposium on Solid Modeling, pp. 285–290 (2004)
6. Porta, J.M., Ros, L., Thomas, F., Corcho, F., Cantó, J., Pérez, J.J.: Complete maps of molecular-loop conformational spaces. Journal of Computational Chemistry 28(13), 2170–2189 (2007)
7. Sippl, M.J., Scheraga, H.A.: Cayley-menger coordinates. Proc. Natl. Acad. Sci. USA 83, 2283 (1986)
8. Sitharam, M., Peters, J., Zhou, Y.: Optimized parametrization of systems of incidences between rigid bodies. Journal of Symbolic Computation 45(4), 481–498 (2010)
9. Liet, T.Y., Lee, T.L., Tsai, C.H.: Hom4ps-2.0: a software package for solving polynomial systems by the polyhedral homotopy continuation method. Computing 83, 109–133 (2008)
10. Thierry, S.E.B., Schreck, P., Michelucci, D., Fünfzig, C., Génevaux, J.-D.: Extensions of the witness method to characterize under-, over- and well-constrained geometric constraint systems. Computer-Aided Design 43(10), 1234–1249 (2011)
11. Thomas, F., Ros, L.: Revisiting trilateration for robot localization. IEEE Transactions on Robotics 21(1), 93–101 (2005)
12. Lin, Q., Gao, X.-S., Zhang, G.-F.: A c-tree decomposition algorithm for 2D and 3D geometric constraint solving. Computer-Aided Design 38(1), 1–13 (2006)
13. Yang, L.: Solving geometric constraints with distance-based global coordinate system. In: Proceedings of the Workshop on Geometric Constraint Solving, Beijing, China (2003), http://www.mmrc.iss.ac.cn/~ascm/ascm03/
14. Yang, L.: Distance coordinates used in geometric constraint solving. In: Winkler, F. (ed.) ADG 2002. LNCS (LNAI), vol. 2930, pp. 216–229. Springer, Heidelberg (2004)

Formal Proof in Coq and Derivation of an Imperative Program to Compute Convex Hulls*

Christophe Brun, Jean-François Dufourd, and Nicolas Magaud

ICube, UMR 7357 University of Strasbourg - CNRS
300 bd Sébastien Brant - BP 10413 - 67412 Illkirch Cedex, France

Abstract. This article deals with a method to build programs in computational geometry from their specifications. It focuses on a case study namely computing incrementally the convex hull of a set of points in the plane using hypermaps. Our program to compute convex hulls is specified and proved correct using the Coq proof assistant. It performs a recursive traversal of the existing convex hull to compute the new hull each time a new point is inserted. This requires using well-founded recursion in Coq. A concrete implementation in Ocaml is then automatically extracted and an efficient C++ program is derived (by hand) from the specification.

1 Introduction

This paper reports on a formal case study in computational geometry on a well-known problem: computing incrementally the convex hull of a finite set of planar points. This fits in a broader project aimed at surveying geometric modeling and computational geometry to improve the programming techniques and ensure the algorithms correctness. Our work is developed in the interactive *Coq* proof assistant [8, 2], which is based on a higher-order intuitionistic logical framework designed to formalize and prove mathematical properties. In addition, we use a generic topology-based approach to specify our algorithms using *hypermaps*.

A (two-dimensional) hypermap is a simple algebraic structure which allows us to model surface subdivisions (into vertices, edges and faces) and to distinguish between their topological and geometric aspects. For years, we have formally described hypermaps to prove topological properties of surfaces [12]. We define hypermaps inductively, which simplifies the construction of operations and proofs. The convex hull computation is a classical planar problem which is not only rich enough to highlight many interesting problems in topology and geometry, but also simple enough to reveal them easily and completely.

The geometric aspects we consider through an *embedding* of the hypermaps in the plane are particularly simple but fundamental in computational geometry. The embedding maps subdivision vertices into points, edges into line segments and faces are represented as polygonal frontiers. However, we also need an orientation predicate to determine whether three points are enumerated clockwise or counter-clockwise. Such a predicate can be defined using real numbers from

---

* This work was supported by the french ANR project Galapagos (2008-2011).

T. Ida and J. Fleuriot (Eds.): ADG 2012, LNAI 7993, pp. 71–88, 2013.
© Springer-Verlag Berlin Heidelberg 2013

the Coq library [8] by computing a determinant and checking its sign. We then simply check that this predicate verifies Knuth's axiom system for orientation. This allows us to isolate the numerical computations (including numerical accuracy issues) from the rest of our study. Investigating these issues is intensively studied by several other researchers, e.g. [18]. We also assume no three points are collinear. This makes our description clearer, thus focusing on the core of our interests, i.e. algorithms and data structures. However, extending our description to the general case would be straightforward but require a lot of time.

A first experiment using structural induction to do the traversal of the hypermap is presented in [6]. At each step, it tests the hypermap elements in a random order but requires to use a copy of the initial hypermap to carry out the orientation tests, which can be costly w.r.t. memory consumption. This approach was quite exploratory and the way the Coq specification was written makes it unlikely to be able to derive it into a realistic imperative program.

In this present paper, we investigate how to deal with a program which is closer to the usual implementation of the incremental algorithm to compute convex hulls, i.e. a program that proceeds by traversing the already-existing hull through its edges instead of investigating darts at random. Although it requires using well-founded induction and thus is somehow technical because of Coq specificities, this leads to a much more natural description than in our previous experiment [6].

Thanks to higher-level operations Merge and Split [13], handling hypermaps also becomes simpler and more accessible to non-specialists of both inductive specifications and hypermaps. This allows to derive more ptopologic properties in the result, including numbering the connected components and the faces of the final subdivision which both were not achieved before [6]. This specification eventually leads to a realistic imperative implementation in C++, derived from the Coq specification, which is integrated in the library CGoGN [17], a result which was not considered achievable in [6].

2   Related Work

Convex Hull and Subdivisions. The convex hull has several definitions and construction methods, such as the incremental algorithm, Jarvis' march, Graham's scan or the divide and conquer approach. It is a fairly simple subdivision of the plane (roughly by a polygon) which still combines topological and geometric aspects. General subdivisions of varieties − into edges, vertices, faces, etc. − have been studied extensively (see [14, 10]) and combinatorially described in [9, 20]. Among these descriptions of subdivisions, *hypermaps* are one of the most general ones, homogeneous in all the dimensions, and easy to formalize algebraically. We shall work with this notion in dimension two but constrain it, at a later time, into *combinatorial oriented maps*, in short *maps*, which is exactly what is required for our study of convex hulls. Maps led to several implementations in geometric modelling, e.g. in the libraries CGAL [15] and CGoGN [17]. We need to embed hypermaps (and maps) into the oriented Euclidian plane to be able to formalize what a convex hull is. As usual, our embedding consists in mapping vertices into points of the plane, all other objects being obtained by linearization. We thus rely on the axiom system for geometric computation and orientation proposed by Knuth [19]. This axiom system, based on properties of triples of points in the plane, allows us to isolate numerical accuracy issues in computations and let us focus on the logical tests required in the algorithms. This approach is well-suited to carry out formal proofs of correctness of the considered algorithms.

Formal Proofs in Computational Geometry. Formal proofs in computational geometry, especially focusing on convex hull algorithms have been carried out in Coq [23] and in Isabelle/HOL [21]. Both use Knuth's axiom system but none of the above-mentioned works relies on any topological structure. However, hypermaps have been used highly successfully to model planar subdivisions in the formalization and proof of the four-color theorem in Coq by Gonthier et al. [16]. Their specification approach as well as the proof techniques (using reflection in Coq) are fairly different from the methodology we follow in this paper. The formal proofs we carry out in Coq follow the tradition of algebraic specifications with constructors as studied in [4]. At Strasbourg University, the library specifying hypermaps, onto which our present work is built, was successfully used to prove some significant results in topology [12]. It was also used to carry out a formal proof of correctness of a functional algorithm of image segmentation and to develop a time-optimal C-program [11]. Recent and on-going works include specifying and proving correct simple algorithms to compute convex hulls [6] and Delaunay triangulations [13]. In addition, we already quoted the improvements (Section 1) provided by this new version of the algorithm and proof compared to the one presented in [6].

An alternative approach, based on certificates, is investigated in [1] for some algorithms about graphs of the LEDA library, which is a platform for combinatorial and geometric computing.

3   Hypermaps

We mathematically define hypermaps and cells and then specify them in Coq.

Definition 1 (Hypermap). *(1) A (two-dimensional) hypermap is an algebraic structure $M = (D, \alpha_0, \alpha_1)$, where D is a finite set, the elements of which are called* darts, *and where α_0, α_1 are permutations on D.*
(2) When α_0 is an involution on D (i.e. $\forall x \in D$, $\alpha_0(\alpha_0(x)) = x$), then M is called a combinatorial oriented map − *in short a* map.

In this framework, 0 and 1 symbolize the two *dimensions*. The topological cells of a hypermap can be combinatorially defined through the classical notion of orbit. Let f be a permutation of darts of a hypermap. The *orbit* of x for f, noted $\langle f \rangle(x)$, is the set of darts accessible from x by iteration of f. We could

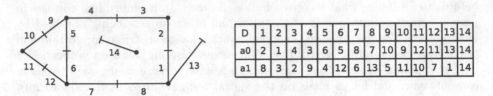

Fig. 1. An example of hypermap (actually a combinatorial oriented map). Intuitively, it represents a subdivision of the plane, into three faces: a rectangle (internal face), a triangle (internal face) which are glued together and an external face with 2 remaining half-segments (13 and 14).

have defined orbits as (circular) sequences as done in the formalization of *tame graphs* in the Flyspeck project [22]. However, it appears more convenient to use sets in this paper.

The orbits for α_0, or 0-*orbits*, are the *edges* of the hypermap and those of α_1, or 1-*orbits*, are its *vertices*. Note that in a map, each edge is composed of at most two darts. *Faces* are the orbits for $\phi = \alpha_1^{-1} \circ \alpha_0^{-1}$. The number of darts of an orbit is called its *degree*. *Connected components* are also defined as usual: the hypermap is considered as a 2-graph equipped with α_0 and α_1, viewed as two binary relations.

An *embedding* of a hypermap is a drawing on a surface where darts are represented by oriented half-segments of curve with the following conventions: (1) the half-segments of all the darts of a given vertex (resp. edge) have the same origin (resp. extremity); (2) the oriented half-segments of an edge, a vertex or a face are traversed in counter-clockwise order when one follows α_0, α_1 or ϕ, respectively; (3) the (open) half-segments do not have any intersection. When the hypermap can be embedded in a plane it is said to be *planar*. In this case, every face which encloses a bounded (resp. unbounded) region on its left is called *internal* (resp. *external*). For more details on embedding and planarity, the reader is refered to [12]. For example, in Fig. 1, a hypermap (in fact a map) $M = (D, \alpha_0, \alpha_1)$ is embedded in the plane with straight half-edges. It has 14 darts, 8 edges (symbolized by small *strokes*), 6 vertices (symbolized by *bullets*), 4 faces and 2 components. For instance, $\langle \alpha_0 \rangle(1) = \{1, 2\}$ is the edge of dart 1 and $\langle \alpha_1 \rangle(1) = \{1, 8, 13\}$ its vertex. We have $\phi(1) = 3$, $\phi(3) = 5$, $\phi(5) = 7$ and $\phi(7) = 1$. The (internal) face of 1 is $\langle \phi \rangle(1) = \{1, 3, 5, 7\}$ and the (external) face of 2 is $\langle \phi \rangle(2) = \{2, 13, 8, 12, 10, 4\}$.

Specification of Free Hypermaps. In our Coq specification, darts, of type dart, are natural numbers and dimensions, of type dim, are zero and one. The hypermaps are first approached by a general notion of *free hypermap*, thanks to a free algebra of terms of inductive type fmap with 3 constructors, V, I and L, respectively for the *empty* (or *void*) hypermap, the *insertion* of a dart, and the *linking* of two darts:

```
Inductive fmap : Set :=
  V : fmap
| I : fmap -> dart -> point -> fmap
| L : fmap -> dim -> dart -> dart -> fmap.
```

For example, a part of the hypermap M of Fig.1, consisting of darts $1, 2, 3$ and 8 is (functionally) represented in Coq by the following term : (L (L (L (I (I (I (I V 1 p1) 2 p2) 3 p3) 8 p8) zero 1 2) one 2 3) one 1 8).

When darts are inserted into a free hypermap, they come together with an embedding point which is a pair of real numbers. That is enough to embed darts on straight half-segments, and edges on straight segments, in the plane. As the reader may see from Fig. 1, some geometrical consistency properties must be enforced. For instance, the points p2 and p3 respectively associated with darts 2 and 3 must be equal. Coq also generates an induction principle on free hypermaps. This principle allows us to prove properties or build functions by induction on the hypermaps.

Specification of Hypermaps. Preconditions written as predicates are introduced for operators I and L to avoid meaningless free maps. The precondition for I states that the dart to be inserted in the free hypermap m must be different from nil $(= 0)$ and from any dart of m. The precondition for L expresses that the darts x and y we want to link together at dimension k in hypermap m are actually already inserted, that x has no k-successor and that y has no k-predecessor w.r.t α_k. The operation α_k is denoted in Coq by cA m k. These preconditions allow us to define an invariant inv_hmap for the hypermap subclass of free maps. It is systematically used in conjunction with fmap. The predicate exd m x expresses that the dart x exists in the hypermap m. Operations cA, cA_1, cF and cF_1 simulate the behavior of the functions α_k, α_k^{-1}, ϕ and ϕ^{-1}. For example, if the hypermap in Fig. 1 is m, exd m 4, cA m one 4 = 9, cA_1 m one 7 = 12, cA m zero 13 = 13, cA_1 m zero 13 = 13. In addition, when the input dart does not belong to the hypermap, we have cA m one 15 = nil and cA_1 m one 15 = nil. Furthermore, we have cF m 1 = 3, cF_1 m 1 = 7.

Merge and Split Functions. Two high-level hypermap operations, not used in the previous experimentation [6], will make easier the writing of our further algorithms. For any dimension k = 0, 1, the operation named Merge merges two k-orbits. To do it, we must choose a dart x in the first one and a dart y in the second one, such that the k-successor of x will be y in the newly-formed orbit (Fig. 2.a). The precondition for Merge requires that x and y do not lie in the same k-orbit.

The operation named Split splits a k-orbit into two pieces with respect to two darts x and y. The precondition for Split is that x and y must be different, but belong to the same k-orbit (Fig. 2.b). We easily prove that, when they satisfy their preconditions, these operations preserve the hypermap invariant inv_hmap. Further topological properties may be considered while proving correct our convex hull algorithm. In addition, invariants dealing with geometry must be defined.

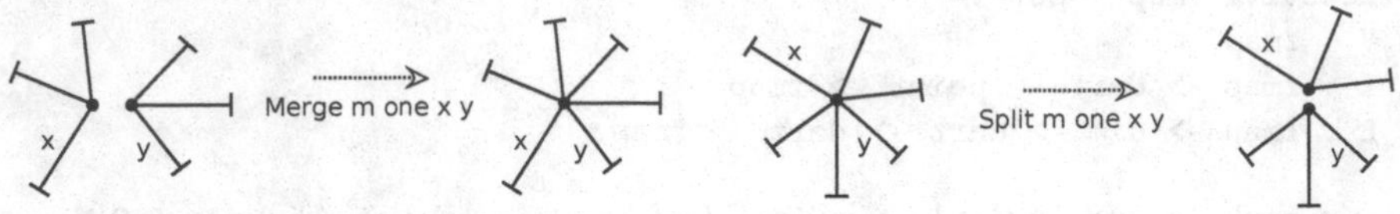

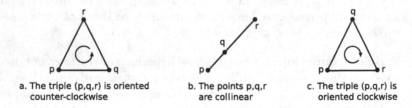

(a) Merging 2 k-orbits   (b) Splitting an orbit

Fig. 2.

4   Convex Hull and Geometric Setting

Computing the convex hull of a set of points requires to determine the orientation of three points of the plane (whose coordinates are real numbers). This is necessary to determine whether a point lies inside or outside a given polygon. We use the orientation predicate $ccw(p, q, r)$ denoted by `ccw p q r` in Coq. It can be defined by $ccw(p, q, r) = det(p, q, r) > 0$ using the determinant:

$$det(p, q, r) = \begin{vmatrix} x_p & y_p & 1 \\ x_q & y_q & 1 \\ x_r & y_r & 1 \end{vmatrix}$$

which also allows us to define a predicate `align p q r` stating that the three points are collinear. In our development, it is established that this implementation is a *model* of Knuth's axioms about orientation [19] which are sufficient to prove all the key geometric properties we need. This approach allows us to deal cleverly with orientation from a logic point of view instead of considering numbers and computations (see [5] for details). Among others, it has been successfully used by Pichardie and Bertot in [23].

a. The triple (p,q,r) is oriented
counter-clockwise

b. The points p,q,r
are collinear

c. The triple (p,q,r) is
oriented clockwise

Fig. 3. The orientation predicate

Convex Hull Definition. In this work, we choose a definition of the convex hull well-suited for our topological hypermap model, for using Knuth's orientation predicate *ccw* and for the incremental algorithm we will study [10, 14]. As our main interest lies in using hypermaps to formalize a convex hull computation algorithm, we assume that points are in *general position*, i.e. *no two points coincide* and *no three ones are collinear*. Other authors, e.g. in [23], study how to relax this restriction using the pertubation method.

Let P be a set of points in the plane.

Definition 2 (Convex hull). *The* convex hull *of P is the polygon T whose vertices t_i, numbered in a counterclockwise order traversal for $i = 1, \ldots, n$ with $n + 1 = 1$, are points of P such that, for each edge $[t_i t_{i+1}]$ of T and for each point p of P different from t_i and t_{i+1}, ccw(t_i, t_{i+1}, p) holds.*

In other words, every point p of P different from t_i and t_{i+1} lies on the left of the oriented line generated by $\overrightarrow{t_i t_{i+1}}$ (Fig. 4).

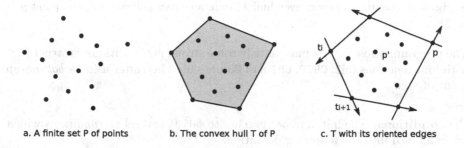

a. A finite set P of points   b. The convex hull T of P   c. T with its oriented edges

Fig. 4. Characterizing a convex hull

Incremental Algorithm. The convex hull of P is built step by step. Each step takes as input the current hull T (the one built with all the already-processed points), a new point p of P and returns a new convex hull T'. Then, either p lies *inside* T and the algorithm moves on to the next step, or it lies *outside* of T and the algorithm will have to remove some edges $[t_i t_{i+1}]$ of T which are *visible* from p — $\neg$ ccw(t_i, t_{i+1}, p) holds —. To build T' it also creates two new edges $[t_l p]$ and $[p t_r]$ connecting p to the *leftmost vertex* t_l and to the *rightmost vertex* t_r (Fig. 5). This corresponds to the usual incremental algorithm (e.g. in [10]). Since we assume that points are in general position, p can never be *on* the already-built polygon, i.e. be equal to a previously-added point or lie on an existing edge.

5 Program Specification

In our specification with hypermaps, the convex hull is an internal face of the current hypermap. To identify it, we first define a new type `mapdart` of pairs (the cartesian product of types is denoted by *), which consists of a current hypermap and a dart of the current convex hull. This definition is mandatory to be able to perform orientation tests in the plane between the inserted point and the existing edges of the current convex hull.

```
Definition mapdart := fmap * dart.
```

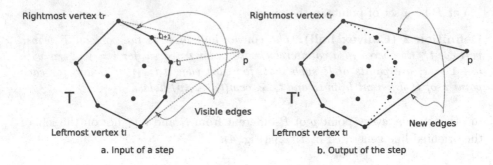

Fig. 5. Computing a new convex hull T' from a convex polygon T and a point p

The program, whose inputs must satisfy some strong preconditions, is structured with four functions CH2, CHID, CHI and CH presented hereafter using a *bottom-up* point of view.

Preconditions. The initial hypermap has to satisfy several *preconditions* which are combined into the following definition:

```
Definition prec_CH (m:fmap) : Prop :=
  inv_hmap m /\ linkless m /\ is_neq_point m /\ is_noalign m.
```

Of course, the free hypermap m must verify the invariant inv_hmap (Section 3), which ensures we have coherent hypermaps. The predicate linkless ensures that it does not contain any links between darts. The predicate is_neq_point states that all darts in the initial hypermap must have different embeddings. The predicate is_noalign ensures three darts with different embeddings can not be embedded into three collinear points. Note that the starting point of our algorithm is actually a set of points, but we model it as a linkless hypermap.

Initialisation Function CH2. It builds a map from two darts as described in Fig. 6. It introduces two new darts, max+1 and max+2, which are linked to the two input darts, x1 and x2, using the function Merge. Note that max is a counter, used to generate fresh new darts to be inserted in the map.

Leftmost and Rightmost Darts. We define two functions search_left and search_right to search resp. for the leftmost and the rightmost darts in a face (m,d) with respect to p, by performing a traversal of the whole set of darts recursively. They both proceed by general recursion, which is achieved in Coq using the keyword Function. A strictly *decreasing measure* has to be provided to Coq to prove the termination. The keyword measure actually expects two arguments, a function to compute the measure and its argument. Informally, these two functions use an integer i which is incremented by 1 at each recursive call (starting from 0). If the degree of the face (that is to say its number of

Fig. 6. A convex hull of two darts built by the CH2 function (edges are slightly curved to make them visible). The embedding of such a convex hull is actually a straigth line connecting the two vertices.

darts), defined in Coq by the predicate `degreef`, is less or equal to `i`, then all darts of the face have been checked and there is no leftmost (resp. righmost) dart. Otherwise, it checks whether the `i`-th successor in the face containing `d` corresponds to the leftmost (resp. rightmost) dart. Functions `search_left` and `search_right` are similar and we only present the code of the first one.

```
Function search_left (m:fmap)(d:dart)(p:point)(i:nat)
  {measure (fun n:nat => (degreef m d) - n) i} :=
  if (le_lt_dec (degreef m d) i) then nil
  else let di := Iter (cF m) i d in
    if (left_dart_dec m di p) then di
    else search_left m d p (i+1).
```

Insertion Function CHID. It computes the convex hull of the current convex polygon denoted by `md := (m,d)` and a new dart `x`. It determines the leftmost dart `l` and the rightmost one `r` in `m` with respect to `x` and its embedding `p`. If `l` is `nil`, then it is proven that `r` is also `nil` and this means `p` lies inside the convex hull. In this case, it simply inserts `x` into the free hypermap. Otherwise, it proceeds in six steps illustrated at Fig. 7.

1. It splits the leftmost dart `l` from its 0-successor `l0 := cA m zero l`.
2. If the dart `l0` is different from `r`, it splits `r` from its 0-predecessor `r_0 := cA_1 m zero r`.
3. It inserts `x` and the *fresh* dart `max` both embedded into `p`.
4. It merges `max` and `x` at dimension `one`.
5. It merges `l` and `max` at dimension `zero` to create a new edge.
6. It merges `x` and `r` at dimension `zero` to close the convex hull.

```
 Definition CHID (md:mapdart)(x:dart)(p:point)(max:dart):mapdart :=
    let m := fst md in let d := snd md in
let l := search_left m d p 0 in
if (eq_dart_dec l nil) then (I m x p, d)
else let r := search_right m l p 0 in
    let l0 := cA m zero l in let r_0 := cA_1 m zero r in
    let m1 := Split m zero l l0 in                    (*1*)
    let m2 := if (eq_dart_dec l0 r) then m1           (*2*)
              else Split m1 zero r_0 r in
    let m3 := (I (I m2 x p) max p) in                 (*3*)
    let m4 := Merge m3 one max x in                   (*4*)
    let m5 := Merge m4 zero l max in                  (*5*)
    let m6 := Merge m5 zero x r in (m6, x).           (*6*)
```

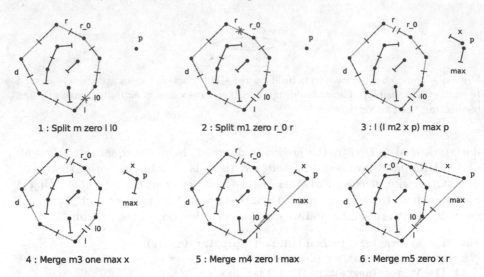

Fig. 7. The six steps of the execution of the insertion function CHID

Recursion Function CHI. It proceeds by recursion and handles the darts of m one by one. At each step, it builds a new convex hull using the insertion function CHID. The free hypermap m corresponds to the set of remaining initial points and md is the already constructed convex hull.

```
Fixpoint CHI (m:fmap)(md:mapdart)(max:dart) {struct m} : mapdart :=
  match m with
  | I m0 x p => (CHI m0 (CHID md x p max) (max+1))
  | _ => md
  end.
```

Main Function CH. It computes the convex hull of a free hypermap m representing the finite set of darts supporting the initial points. If m has a single dart x, it returns a pair (m, x). If it has at least 2 darts, CH builds a first convex polygon with two of these darts using CH2 (Fig. 6) and then calls CHI. To have an exhaustive pattern-matching, we have a default case which actually never happens because of the preconditions and it simply returns a pair formed by the initial set of darts and the dart nil.

```
Definition CH (m:fmap) : mapdart :=
  match m with
  | I V x p => (m,x)
  | I (I m0 x1 p1) x2 p2 =>
    CHI m0 (CH2 x1 p1 x2 p2 (max_dart m))
        ((max_dart m)+3)
  | _ => (m,nil)
  end.
```

This definition of CH is similar to the one appearing in the usual incremental algorithm [10, 14]. However, it is important to note that chains of darts which are inside the polygon remain in the computed hypermap. These useless chains of darts could be removed in a function built on top of CH or each time CHID is executed.

6   Topological Properties

The structure of the proofs, especially for those related to the insertion function CHID, closely follows the structure of the program. Therefore, we successively prove each of the following properties for the six successive hypermaps computed by the algorithm and presented in the previous section. Note that all the proofs are drastically lighter than in our first experiment [6], and that new results about the numbering of connected components and faces are obtained in our present version.

Hypermap Invariant and Initial Darts. The first important theorem we prove is a technical one. Indeed, we have to show that the invariant inv_hmap holds throughout the program, from the initial hypermap m to the final hypermap (CH m). This states that a dart can not be inserted twice in a hypermap and that darts must belong to the hypermap before being linked together (see Section 3).

```
Theorem inv_hmap_CH : forall (m:fmap),
  prec_CH m -> inv_hmap (fst (CH m)).
```

The proof proceeds in 6 steps according to the structure of the program. Each step is fairly easy to prove thanks to the invariant-preserving properties of I, Merge and Split. In the same way, we prove that all the initial darts are preserved, which also entails that all the initial points are preserved.

Involutions without Fixpoint. One of the properties of the convex hull is to be a polygon, which means that each edge and vertex is of degree 2. This property is expressed by the predicate inv_gmap which states that for each dart x of a hypermap m, and for each dimension k, α_k is an involution.

```
Definition inv_gmap (m:fmap) : Prop :=
  forall (k:dim)(x:dart), exd m x -> cA m k (cA m k x) = x.
```

The proof that the convex hull satisfies this property is given by:

```
Theorem inv_gmap_CH : forall (m:fmap),
  prec_CH m -> inv_gmap (fst (CH m)).
```

Another property of the polygon representing the convex hull is that it does not have any fixpoint with respect to (cA m k), k=zero or k=one. This property is ensured by the predicate inv_poly. Thus, for each dart x belonging to the same connected component as d in the hypermap m (i.e. verifying eqc m d x), whatever the dimension k, none of the k-orbits admits a fixpoint.

```
Definition inv_poly (m:fmap)(d:dart) : Prop :=
  forall (k:dim)(x:dart), eqc m d x -> x <> cA m k x.
```

This definition uses the notion of connected components which is more intuitive than the notion of faces. However, note that it is established that if darts x and y belong to the same face, then they both belong to the same connected component. Then we prove the following theorem:

```
Theorem inv_poly_CH : forall (m:fmap),
  prec_CH m -> inv_poly (fst (CH m)) (snd (CH m)).
```

Planarity. We now verify the polygon we build is planar. The definition of the planarity property `planar` from Euler's formula can be found in the proof development and is omitted here.

```
Theorem planar_CH : forall (m:fmap),
  prec_CH m -> planar (fst (CH m)).
```

The proof of this theorem uses several planarity criteria [12] for `Merge` and `Split` and is fairly straightforward.

Connected Components Numbering. Our new approach allows to exactly count the connected components of the subdivision at each step of the process. With the notations of section 5, the following result is proved for each `CHI` application:

- (a) If `l0=r`, the number of connected components remains the same.
- (b) Otherwise, it exactly increases of 1. Indeed, a new dart chain, possibly reduced to a single dart, lies inside the convex hull.

This entails, when the internal chains are eliminated, that the number of connected components is always 1: it is easy to prove that it is 1 for `CH2` and it remains equal to 1 at each `CHI` application. Note that this result is deduced from good synthetical properties of `Split` and `Merge` (see [5] for details).

Faces Numbering. This approach leads to a similar result for the faces of the subdivision. The variation of the face number is exactly proved to be the same as for components. However, when the internal chains are removed, it exactly has the value 2 for `CH2` and keeps this value in `CHI`, which corresponds to the internal and external faces of the convex hull.

7   Geometric Properties

The geometric properties we prove ensure that darts are embedded in the plane in a coherent manner and that the program builds a polygon that is actually convex (see Definition 2).

Embedding, Distinct and Collinear Points. We first prove that darts are embedded correctly with respect to their links: darts x and y which belong to the same vertex (eqv m x y) must have the same embedding whereas darts belonging to the same edge (eqe m x y) must have different embeddings. This is summarized in the following definition is_well_emb:

```
Definition is_well_emb (m:fmap) : Prop :=
   forall (x y : dart), exd m x -> exd m y -> x <> y ->
   let px := fpoint m x in let py := fpoint m y in
   (eqv m x y -> px = py) /\ (eqe m x y -> px <> py).
```

Then we prove the following theorem:

```
Theorem iswellemb : forall (md:mapdart)(x max :dart)(p:point),
   is_well_emb (fst (CHID md x p max)).
```

Finally, one of our initial preconditions, verified throughout the execution of the program states that there are no three collinear points. This property is defined by the predicate is_noalign (whose definition is omitted here) and we prove it holds after CHID:

```
Theorem noalign : forall (md:mapdart)(x max :dart)(p:point),
   is_noalign (fst (CHID md x p max)).
```

Convexity. The central property states the polygon we build is actually convex. The definition is_convex expresses this property: a face (identified by (m, d)) is convex if for each dart x of this face (eqf m d x) and for each dart y whose embedding py is different from the embeddings px of x and px0 of the 0-successor x0 := cA m zero x of x, the triple (px, px0, py) is oriented counter-clockwise:

```
Definition is_convex (m:fmap)(d:dart) : Prop :=
   forall (x:dart)(y:dart), eqf m d x -> exd m y ->
   let px := fpoint m x in let py := fpoint m y in
   let px0 := fpoint m (cA m zero x) in
   px <> py -> px0 <> py -> ccw px px0 py.
```

The theorem convex states that the result of the insertion function CHID verifies the convexity property. It is clear that the property also holds for CH.

```
Theorem convex : forall (md:mapdart)(x max :dart)(p:point),
   is_convex (fst (CHID md x p max)).
```

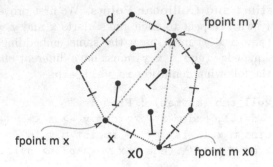

Fig. 8. The convexity property

Convex Hull Property. We put together all the properties required to have a convex hull. This is defined as follows:

```
Definition is_convex_hull (md:mapdart) : Prop :=
  let m := (fst md) in let d := (snd md) in
  inv_hmap m /\ inv_gmap m /\ inv_poly m d /\ planar m /\
  is_well_emb m /\ is_neq_point m /\ is_noalign m /\ is_convex m d.
```

8   Program Derivation in C++

Extraction into OCaml. Coq features an extraction mechanism which can be used to automatically transform our Coq specification into an executable Ocaml program. This allows us to test *in real conditions* how our data structures and algorithm behave in terms of both usability and efficiency before starting any formal proof. The program extracted contains the code of functions CH, CHI, CH2, and CHID. We reuse the graphical interface we presented in [6] to select points of the plane and display the convex hull as a polygonal line (together with some remaining isolated points inside). For lack of space, we do not present the generated Ocaml code which is very close to the Coq specification and can be found in [5]. As such a basic extraction process may raise some efficiency issues especially because of the use of pattern-matching to deal with hypermaps, we also study how to integrate our formally-proved program into a general platform for geometric modeling.

Implementing Hypermaps in C++. We derive a C++ implementation of an optimized program from our specification and integrate it into the library CGoGN (standing for Combinatorial and Geometric mOdeling with Generic N-dimensional Maps) [17] developed in our team to deal with hypermaps. The transformation is performed manually but most operations remain the same and are very close to those initially written in Coq. In the library CGoGN, hypermaps are represented by doubly-linked dart lists inherited from the Standard Template

Library (STL) of C++. A dart in a hypermap is a pointer on a structure with a dart array to describe the topology (i.e. the links of the dart to its predecessors and successors) and, in our case, a pointer to a point structure equipped with a counter recording how many darts are embedded into this point (see Fig. 9).

Hypermaps have the type hmap. We build the C++ datatype mapdart of pairs and implement the corresponding access functions fstmd and sndmd, as well as the pair constructor pairmd.

```
typedef struct mapdartstruct { hmap m; dart d; } mapdart;
```

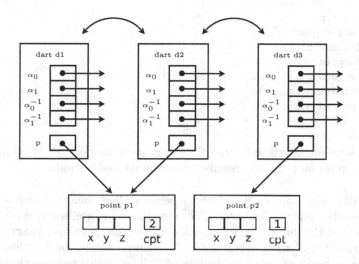

Fig. 9. Representing hypermaps in C++ with CGoGN

Computing the Convex Hull. We easily implement functions from our initial library on hypermaps, namely the atomic operations V, I and L and the higher-level ones Merge and Split. Note that the latter are programmed with side effects on the hypermap m and simply rewrite the already-existing links. We rewrite the Coq specification of the two functions search_left and search_right into two similar recursive C++ programs.

The main functions CH and CHI do not proceed by pattern matching on the initial hypermap m but require to extract and then remove explicitly each dart using CCoGN functions. Finally, the insertion function CHID is programmed as follows. To create a dart in the hypermap, we simply use the function gendart which returns an identifier (a pointer), and the result of gendart is used by the function I which inserts the dart together with its embedding.

```
mapdart CHID (mapdart md, const point p) {
  hmap m = fstmd(md); dart d = sndmd(md);
  dart l = search_left(m,d,p,0);
  dart r = search_right(m,cA(m,zero,d),p,0);
  if (l==m->nil()) {
    dart x = gendart(m);
    m = I(m,x,p);
    md = pairmd(m,d);
  } else {
    dart l0 = cA(m,zero,l); dart r_0 = cA_1(m,zero,r);
    m = Split(m,zero,l,l0);
    if (l0!=r) m = Split(m,zero,r_0,r);
    dart x = gendart(m);
    m = I(m,x,p);
    dart max = gendart(m);
    m = I(m,max,p);
    m = Merge(m,one,max,x);
    m = Merge(m,zero,l,max);
    m = Merge(m,zero,x,r);
    md = pairmd(m,x); }
  return md; }
```

We also design an interface for the C++ program we derived. It allows us to test our program and also to visualize their inputs and outputs.

Complexity. The complexity of the insertion C++ function CHI as implemented from its Coq specification is $O(n^2)$ in the worst case, where n is the number of points of the initial set. In the code of search_left and search_right, the calls to Iter (cF m) i d is replaced by a simple call to the function ϕ which returns the successor of a dart in the face of d. This avoids recomputing the first i successors of d at each step. Consequently, the complexity of the algorithm is $O(n^2)$ as in any implementation of this kind of incremental algorithm. Indeed, all algorithms which are more efficient to compute convex hulls are not incremental and do require to know about *all* the involved vertices before starting the computation. After this experiment, we could more easily revisit and implement various computational gemetry algorithms based on hypermaps and the orientation predicate and achieve the same complexity as algorithms usually presented in the litterature.

9   Conclusion

The algorithm we specified proceeds by a traversal of the current convex hull when a new point is inserted. It is closer to the usual implementations [14, 10], than the one we presented in [6]. We proved that the algorithm is correct by establishing in Coq its topological and geometric properties. We also automatically extracted our specification into a prototype Ocaml program and derived an implementation in C++ using the library CGoGN on hypermaps.

Compared to the approach followed in [6], this work presents a higher-level specification and more compact proofs, especially thanks to `Merge` and `Split`. In addition, we follow the usual behavior of the incremental algorithm and prove more properties than in our first attempt. Proofs about numbering of faces and connected components, especially the proofs they increase by at most one with each insertion of a dart, are successfully carried out. Finally, we could consider removing the dangling chains of darts lying inside the convex hull and prove we have only one connected component and two faces (internal and external) at the end, which are crucial properties in topology-based geometric modeling. Our specification, which relies on the hypermap library developped in Strasbourg, fits in about 2500 lines and the proofs require around 6000. Note that the size of this basic hypermaps library grew extensively since our first experiment [6] and now contains several higher-level operations as well as proofs of their properties. Proofs are shorted and easier to write down thanks to the use of the operations `Merge` and `Split`.

The program derived in C++ remains very close to the specification in Coq. A further research direction could be to carry out a formal proof of the implementation using tools such as Why and Frama-C [7] as was done by one of the authors for integer arithmetics [3]. Conversely, it would even nicer if we could directly derive a C program from the Coq specification. Indeed, our overall reseach direction is to derive programs from their specifications rather than proving correct already existing programs (as people do with Hoare logic). This remains a difficult challenge, especially because of simple data structures (such as lists) are mixed with pointers, which leads to more complex data structures which are hard to handle in frameworks such as Why and Frama-C. Finally, other computational geometry algorithms [13] are currently revisited using our formal approach with Coq and hypermaps. Furthermore, a great challenge would be to investigate the third dimension, by considering surfacic convex hulls as well as volumic subdivisions − into edges, vertices, faces and volumes −, for instance for tetrahedric Delaunay diagrams.

References

[1] Alkassar, E., Böhme, S., Mehlhorn, K., Rizkallah, C.: Verification of Certifying Computations. In: Gopalakrishnan, G., Qadeer, S. (eds.) CAV 2011. LNCS, vol. 6806, pp. 67–82. Springer, Heidelberg (2011)

[2] Bertot, Y., Castéran, P.: Interactive Theorem Proving and Program Development, Coq'Art: The Calculus of Inductive Constructions. Texts in Theoretical Computer Science, An EATCS Series. Springer (2004)

[3] Bertot, Y., Magaud, N., Zimmermann, P.: A Proof of GMP Square Root. Journal of Automated Reasoning 29(3-4), 225–252 (2002)

[4] Bertrand, Y., Dufourd, J.-F., Françon, J., Lienhardt, P.: Algebraic Specification and Development in Geometric Modeling. In: Gaudel, M.-C., Jouannaud, J.-P. (eds.) CAAP 1993, FASE 1993, and TAPSOFT 1993. LNCS, vol. 668, pp. 75–89. Springer, Heidelberg (1993)

[5] Brun, C., Dufourd, J.-F., Magaud, N.: Formal Proof of the Incremental Convex Hull Algorithm, http://galapagos.gforge.inria.fr

[6] Brun, C., Dufourd, J.-F., Magaud, N.: Designing and Proving Correct a Convex Hull Algorithm with Hypermaps in Coq. Computational Geometry, Theory and Applications 45(8), 436–457 (2012)

[7] CEA-INRIA. The Frama-C Software Verification Toolkit, http://frama-c.cea.fr

[8] Coq Development Team. The Coq Proof Assistant - Reference Manual and Library, http://coq.inria.fr/

[9] Cori, R.: Un code pour les graphes planaires et ses applications. Astérisque, vol. 27. Socieété mathématique de France (1970)

[10] de Berg, M., Cheong, O., van Kreveld, M., Overmars, M.: Computational Geometry, Algorithms and Applications. Springer (2008)

[11] Dufourd, J.-F.: Design and Formal Proof of a New Optimal Image Segmentation Program with Hypermaps. Pattern Recognition 40(11), 2974–2993 (2007)

[12] Dufourd, J.-F.: An Intuitionistic Proof of a Discrete Form of the Jordan Curve Theorem Formalized in Coq with Combinatorial Hypermaps. Journal of Automated Reasoning 43(1), 19–51 (2009)

[13] Dufourd, J.-F., Bertot, Y.: Formal Study of Plane Delaunay Triangulation. In: Kaufmann, M., Paulson, L.C. (eds.) ITP 2010. LNCS, vol. 6172, pp. 211–226. Springer, Heidelberg (2010)

[14] Edelsbrunner, H.: Algorithms in Combinatorial Geometry. Monographs in Theoretical Computer Science, An EATCS Series. Springer (1987)

[15] Flato, E., Halperin, D., Hanniel, I., Nechushtan, O.: The Design and Implementation of Planar Maps in CGAL. In: Vitter, J.S., Zaroliagis, C.D. (eds.) WAE 1999. LNCS, vol. 1668, pp. 154–168. Springer, Heidelberg (1999)

[16] Gonthier, G.: Formal Proof - The Four-Colour Theorem. Notices of the AMS 55(11), 1382–1393 (2008)

[17] IGG Team. CGoGN: Combinatorial and Geometric mOdeling with Generic N-dimensional Maps, https://iggservis.u-strasbg.fr/CGoGN/

[18] Kettner, L., Mehlhorn, K., Pion, S., Schirra, S., Yap, C.-K.: Classroom Examples of Robustness Problems in Geometric Computations. Computational Geometry 40(1), 61–78 (2008)

[19] Knuth, D.E.: Axioms and Hulls. LNCS, vol. 606. Springer, Heidelberg (1992)

[20] Lienhardt, P.: Topological Models for Boundary Representation: a Comparison with n-Dimensional Generalized Maps. Computer-Aided Design 23, 59–82 (1991)

[21] Meikle, L.I., Fleuriot, J.D.: Mechanical Theorem Proving in Computational Geometry. In: Hong, H., Wang, D. (eds.) ADG 2004. LNCS (LNAI), vol. 3763, pp. 1–18. Springer, Heidelberg (2006)

[22] Nipkow, T., Bauer, G., Schultz, P.: Flyspeck I: Tame graphs. In: Furbach, U., Shankar, N. (eds.) IJCAR 2006. LNCS (LNAI), vol. 4130, pp. 21–35. Springer, Heidelberg (2006)

[23] Pichardie, D., Bertot, Y.: Formalizing Convex Hull Algorithms. In: Boulton, R.J., Jackson, P.B. (eds.) TPHOLs 2001. LNCS, vol. 2152, pp. 346–361. Springer, Heidelberg (2001)

From Tarski to Hilbert

Gabriel Braun and Julien Narboux

ICube, UMR 7357 University of Strasbourg - CNRS, France

Abstract. In this paper, we report on the formal proof that Hilbert's axiom system can be derived from Tarski's system. For this purpose we mechanized the proofs of the first twelve chapters of Schwabäuser, Szmielew and Tarski's book: *Metamathematische Methoden in der Geometrie*. The proofs are checked formally within classical logic using the Coq proof assistant. The goal of this development is to provide clear foundations for other formalizations of geometry and implementations of decision procedures.

1   Introduction

Euclid is considered as the pioneer of the axiomatic method. In the *Elements*, starting from a small number of self-evident truths, called postulates or common notions, he derives by purely logical rules most of the geometrical facts that were discovered in the two or three centuries before him. But upon a closer reading of Euclid's *Elements*, we find that he does not adhere as strictly as he should to the axiomatic method. Indeed, at some steps in some proofs he uses a method of "superposition of triangles". This kind of justification cannot be derived from his set of postulates[1].

In 1899, in *die Grundlagen der Geometrie*, Hilbert described a more formal approach and proposed a new axiom system to fill the gaps in Euclid's system.

Recently, the task consisting in mechanizing Hilbert's *Grundlagen der Geometrie* has been partially achieved. A first formalization using the Coq proof assistant [2] was proposed by Christophe Dehlinger, Jean-François Dufourd and Pascal Schreck [3]. This first approach was realized in an intuitionist setting, and concluded that the decidability of point equality and collinearity is necessary to check Hilbert's proofs. Another formalization using the Isabelle/Isar proof assistant [4] was performed by Jacques Fleuriot and Laura Meikle [5] and then extended by Phil Scott [6]. These formalizations have concluded that, even if Hilbert has done some pioneering work about formal systems, his proofs are in fact not fully formal, in particular degenerate cases are often implicit in the presentation of Hilbert. The proofs can be made more rigorous by machine assistance. Indeed, in the different editions of *die Grundlagen der Geometrie* the axioms were changed, but the proofs were not always changed accordingly. This obviously resulted in some inconsistencies. The use of a proof assistant solves

---

[1] Recently, Jeremy Avigad and Edward Dean and John Mumma have shown that it is possible to define a formal system to model the proofs of Euclid's Elements [1].

T. Ida and J. Fleuriot (Eds.): ADG 2012, LNAI 7993, pp. 89–109, 2013.
© Springer-Verlag Berlin Heidelberg 2013

this problem: when an axiom is changed it is easy to check if the proofs are still valid. In [7], Phil Scott and Jacques Fleuriot proposed a tool to write readable formalised proof-scripts that correspond to Hilbert's prose arguments. In [6], Phil Scott remarks that giving a formal definition for Hilbert's axioms is a matter of interpretation and it is an error prone process.

In the early 60s, Wanda Szmielew and Alfred Tarski started the project of a treatise about the foundations of geometry based on another axiom system for geometry designed by Tarski in the 20s[2]. A systematic development of Euclidean geometry was supposed to constitute the first part but the early death of Wanda Szmielew put an end to this project. Finally, Wolfram Schwabhäuser continued the project of Wanda Szmielew and Alfred Tarski. He published the treatise in 1983 in German: *Metamathematische Methoden in der Geometrie* [9]. In [10], Art Quaife used a general purpose theorem prover to automate the proof of some lemmas in Tarski's geometry, but the lemmas which can be solved using this technique are some simple lemmas which can be proved within Coq using the `auto` tactic. The axiom system of Tarski is quite simple and has good meta-theoretical properties. Tarski's axiomatization has no primitive objects other than points. This allows us to change the dimension of the geometric space without changing the language of the theory (whereas in Hilbert's system one needs the notion of "plane"). Some axioms provide a means to define the lower and upper dimension of the geometric space. Gupta proved the axioms independent [11], except the axiom of Pasch and the reflexivity of congruence (which remain open problems).

In this paper we describe our formalization of the first twelve chapters of the book of Wolfram Schwabhäuser, Wanda Szmielew and Alfred Tarski in the Coq proof assistant. Then we answer an open question from [5]: Hilbert's axioms can be derived from Tarski's axioms and we give a mechanized proof. Alfred Tarski worked on the axiomatization and meta-mathematics of Euclidean geometry from 1926 until his death in 1983. Several axiom systems were produced by Tarski and his students. In this formalization, we use the version presented in [9].

Formalization of geometry has already been partially addressed by the community. Frédérique Guilhot has realized a large Coq development about Euclidean geometry following a presentation suitable for use in French high-school [12] and Tuan-Minh Pham has further improved this development [13]. We have presented the formalization and implementation in the Coq proof assistant of the area decision procedure of Chou, Gao and Zhang [14–17] and of Wu's method [18, 19].

Formalizing geometry in a proof assistant has not only the advantage of providing a very high level of confidence in the proof generated, it also permits us to insert purely geometric arguments within other kind of proofs such as, for instance, proof of correctness of programs or proofs by induction. But for the time being most of the formal developments we have cited are distinct and as they do not use the same axiomatic system, they cannot be combined. In [20], we have

---

[2] These historical pieces of information are taken from the introduction of the publication by Givant in 1999 [8] of a letter from Tarski to Schwabhäuser (1978).

shown how to prove the axioms of the area method within the formalization of geometry by Guilhot and Pham.

The goal of our mechanization is to do another step toward the merging of all these developments. We aim at providing very clear foundations for other formalizations of geometry and implementations of decision procedures.

We will first describe the axiom system of Tarski, its formalization within the Coq proof assistant (Section 2.1). As our other Coq developments about geometry, we limit ourselves to 2-dimensional geometry. Then we give a quick overview of the formalization (Section 2.2). To show the difficulty of the task, we will give the proof of one of the non trivial lemmas which was not proved by Tarski and his co-authors although they are used implicitly. Then we describe Hilbert's axiom system and its formalization in Coq (Section 3). Finally, we describe how we can define the concepts of Hilbert's axiom system and prove the axioms within Tarski's system (Section 4).

2 Tarski's Geometry

2.1 Tarski's Axiom System

In this section we describe the axiom system we used in the formalization. Further discussion about the history of this axiom system and the different versions can be found in [21]. The axioms can be expressed using first order logic and two predicates. Note that the original theory of Tarski assumes first order logic. Our formalization is performed in a higher order logic setting (the calculus of constructions), hence, the language allowed in the statements and proofs makes the theory more expressible. The meta-theoretical results of Tarski may not apply to our formalization.

betweenness. The ternary *betweenness* predicate $\beta\ ABC$ informally states that B lies on the line AC between A and C. The relation holds if $A = B$ or $B = C$.

congruence. The quaternary *congruence* predicate $AB \equiv CD$ informally means that the distance from A to B is equal to the distance from C to D.

Note that in Tarski's geometry, only points are primitive objects. In particular, lines are *defined* by two distinct points whereas in Hilbert's axiom system lines and planes are *primitive objects*. Figure 1 provides the list of axioms that we used in our formalization.

The formalization of this axiom system in Coq is straightforward (Fig. 2). Coq is a proof assistant which allows the expression of mathematical assertions and mechanically checks proofs of these assertions. It is based on the theory of inductive constructions and the Curry-Howard correspondence. x:T can either mean that the program x is of type T or that x is a proof of T. = denotes Leibniz equality :Given any x and y, $x = y$ if, given any predicate P, $P(x)$ if and only if $P(y)$. A<>B means $A \neq B$. We list here the logical notations used in this paper:

$$\text{Identity } \beta\,A\,B\,A \Rightarrow (A = B)$$
$$\text{Pseudo-Transitivity } AB \equiv CD \land AB \equiv EF \Rightarrow CD \equiv EF$$
$$\text{Symmetry } AB \equiv BA$$
$$\text{Identity } AB \equiv CC \Rightarrow A = B$$
$$\text{Pasch } \beta\,A\,P\,C \land \beta\,B\,Q\,C \Rightarrow \exists X, \beta\,P\,X\,B \land \beta\,Q\,X\,A$$
$$\text{Euclid } \exists XY, \beta\,A\,D\,T \land \beta\,B\,D\,C \land A \neq D \Rightarrow$$
$$\beta\,A\,B\,X \land \beta\,A\,C\,Y \land \beta\,X\,T\,Y$$
$$AB \equiv A'B' \land BC \equiv B'C' \land$$
$$\text{5 segments } AD \equiv A'D' \land BD \equiv B'D' \land$$
$$\beta\,A\,B\,C \land \beta\,A'\,B'\,C' \land A \neq B \Rightarrow CD \equiv C'D'$$
$$\text{Construction } \exists E, \beta\,A\,B\,E \land BE \equiv CD$$
$$\text{Lower Dimension } \exists ABC, \neg\beta\,A\,B\,C \land \neg\beta\,B\,C\,A \land \neg\beta\,C\,A\,B$$
$$\text{Upper Dimension } AP \equiv AQ \land BP \equiv BQ \land CP \equiv CQ \land P \neq Q$$
$$\Rightarrow \beta\,A\,B\,C \lor \beta\,B\,C\,A \lor \beta\,C\,A\,B$$
$$\text{Continuity } \forall XY, (\exists A, (\forall xy, x \in X \land y \in Y \Rightarrow \beta\,A\,x\,y)) \Rightarrow$$
$$\exists B, (\forall xy, x \in X \Rightarrow y \in Y \Rightarrow \beta\,x\,B\,y).$$

Fig. 1. Tarski's axiom system

Connective:	$\neg$	$\land$	$\lor$	$\Rightarrow$	$\Leftrightarrow$	$\forall$	$\exists$
Coq notation:	~	/\	\/	->	<->	forall	exists

We use the Coq type class mechanism [22] to capture the axiom system. Internally the type class system is based on records containing types, functions and properties about them. Note that we know that this system of axioms has

```
Class Tarski := {
 Tpoint : Type;
 Bet  : Tpoint -> Tpoint -> Tpoint -> Prop;
 Cong : Tpoint -> Tpoint -> Tpoint -> Tpoint -> Prop;
 between_identity : forall A B, Bet A B A -> A=B;
 cong_pseudo_reflexivity : forall A B : Tpoint, Cong A B B A;
 cong_identity : forall A B C : Tpoint, Cong A B C C -> A = B;
 cong_inner_transitivity : forall A B C D E F : Tpoint,
   Cong A B C D -> Cong A B E F -> Cong C D E F;
 inner_pasch : forall A B C P Q : Tpoint,
   Bet A P C -> Bet B Q C -> exists x, Bet P x B /\ Bet Q x A;
 euclid : forall A B C D T : Tpoint,
   Bet A D T -> Bet B D C -> A<>D ->
   exists x, exists y, Bet A B x /\ Bet A C y /\ Bet x T y;
 five_segments : forall A A' B B' C C' D D' : Tpoint,
   Cong A B A' B' -> Cong B C B' C' -> Cong A D A' D' -> Cong B D B' D' ->
   Bet A B C -> Bet A' B' C' -> A <> B -> Cong C D C' D';
 segment_construction : forall A B C D : Tpoint,
   exists E : Tpoint, Bet A B E /\ Cong B E C D;
 lower_dim : exists A, exists B, exists C, ~ (Bet A B C \/ Bet B C A \/ Bet C A B);
 upper_dim : forall A B C P Q : Tpoint,
   P <> Q -> Cong A P A Q -> Cong B P B Q -> Cong C P C Q ->
   (Bet A B C \/ Bet B C A \/ Bet C A B)
}
```

Fig. 2. Tarski's axiom system as a Coq type class

a model: Tuan Minh Pham has shown that these axioms can be derived from Guilhot's development using an axiom system based on mass points [13].

2.2 Overview of the Formalization of the Book

The formalization closely follows the book [9]. But many lemmas are used implicitly in the proofs and are not stated by the original authors. We first give a quick overview of the different notions introduced in the formal development. Then we provide as an example a proof of a lemma which was not given by the original authors. This lemma is not needed to derive Hilbert's axioms but it is a key lemma for the part of our library about angles. The proof of this lemma represents roughly 100 lines of the 23000 lines of proof of the whole Coq development. About 65% of the twelve chapters are used for the proof of Hilbert's axioms.

We provide some statistics about the different chapters in Table 1.

Table 1. Statistics about the development

Chapter	Number of lemmas	Number of lines of specification	Number of lines of proof	Lines per lemma
Betweeness properties	16	69	111	6.93
Congruence properties	16	54	116	7,25
Properties of betweeness and congruence	19	151	183	9.63
Order relation over pair of points	17	88	340	20
The ternary relation out	22	103	426	19,36
Property of the midpoint	21	101	758	36,09
Orthogonality lemmas	77	191	2412	141,88
Position of two points relatively to a line	37	145	2333	63,05
Orthogonal symmetry	44	173	2712	61,63
Properties about angles	187	433	10612	56,74
Parallelism	68	163	3560	52,35
Total	**524**	**1671**	**23563**	**45**

The Different Concepts Involved in Tarski's Geometry. We followed closely the order given by Tarski to introduce the different concepts of geometry and their associated lemmas.

Chapter 2: betweeness properties
Chapter 3: congruence properties
Chapter 4: Properties of Betweeness and Congruence. This chapter introduces the definition of the concept of collinearity:

Definition 1 (collinearity). *To assert that three points A, B and C are collinear we note: Col A B C*

$$Col\ A\ B\ C := \beta\ A\ B\ C \vee \beta\ A\ C\ B \vee \beta\ B\ A\ C$$

Chapter 5: Order Relation over Pair of Points. The   relation   *bet_le* between two pair of points formalizes the fact that the distance of the first pair of points is less than the distance between the second pair of points:

Definition 2 (*bet_le*)

$$bet\_le\ A\ B\ C\ D := \exists y, \beta\ C\ y\ D \wedge AB \equiv Cy$$

Chapter 6: The Ternary Relation Out. $Out\ P\ A\ B$ means that P, A and B lies on the same line, but P is not between A and B:

Definition 3 (*Out*)

$$Out\ P\ A\ B := A \neq P \wedge B \neq P \wedge (\beta\ P\ A\ B \vee \beta\ P\ B\ A)$$

Chapter 7: Property of the Midpoint. This chapter provides a definition for midpoint but the existence of the midpoint will be proved only in Chapter 8.

Definition 4 (midpoint)

$$is\_midpoint\ M\ A\ B := \beta\ A\ M\ B \wedge AM \equiv BM$$

Chapter 8: Orthogonality Lemmas. To work on orthogonality, Tarski introduces three relations:

Definition 5 (*Per*)

$$Per\ A\ B\ C := \exists C', is\_midpoint\ B\ C\ C' \wedge AC \equiv AC'$$

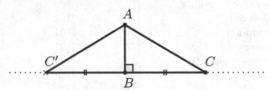

Definition 6 (*Perp_in*)

$$Perp\_in\ X\ A\ B\ C\ D := A \neq B \wedge C \neq D \wedge Col\ X\ A\ B \wedge Col\ X\ C\ D \wedge$$
$$(\forall U\ V, Col\ U\ A\ B \Rightarrow Col\ V\ C\ D \Rightarrow Per\ U\ X\ V)$$

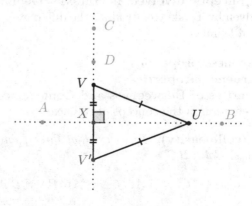

Finally, the relation Perp which we denote by the infix $\perp$:

Definition 7 (*Perp*)

$$AB \perp CD := \exists X, Perp\_in\, X\, A\, B\, C\, D$$

Chapter 9: Position of Two Points Relatively to a Line. In this chapter, Tarski introduces two predicates to assert the fact that two points which do not belong to a line are either on the same side, or on opposite sides of the line.

Definition 8 (opposite sides). *Given a line l defined by two distinct points A and B, two points X and Y not on l, are on opposite sides of l is written: $A\!-\!\frac{X}{Y}\!-\!B$*

$$A\frac{X}{Y}B := \exists T, Col\, A\, B\, T \wedge \beta\, X\, T\, Y$$

Definition 9 (same side). *Let l be a line defined by two distinct points A and B. Two points X and Y not on l, are on the same side of l is written: $A\!-\!\frac{}{XY}\!-\!B$*

$$A\frac{}{XY}B := \exists Z, A\frac{X}{Z}B \wedge A\frac{Y}{Z}B$$

Chapter 10: Orthogonal Symmetry. The predicate *is_image* allows us to assert that two points are symmetric. Let l be a line defined by two distinct points A and B. Two points P and P' are symmetric points relatively to the line l means:

Definition 10 (*is_image*)

$$is\_image\, P\, P'\, A\, B :=$$
$$(\exists X, is\_midpoint\, X\, P\, P' \wedge Col\, A\, B\, X) \wedge (AB \perp PP' \vee P = P')$$

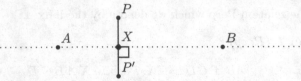

Chapter 11: Properties about Angles. In this chapter, Tarski gives a definition of angle congruence using the similarity of triangles:

Definition 11 (angle congruence)

$$\angle ABC \cong \angle DEF := A \neq B \Rightarrow B \neq C \Rightarrow D \neq E \Rightarrow F \neq F \Rightarrow$$

$$\exists A', \exists C', \exists D', \exists F' \begin{cases} \beta\,B\,A\,A' \wedge AA' \equiv ED \wedge \\ \beta\,B\,C\,C' \wedge CC' \equiv EF \wedge \\ \beta\,E\,D\,D' \wedge DD' \equiv BA \wedge \\ \beta\,E\,F\,F' \wedge FF' \equiv BC \wedge \\ A'C' \equiv D'F' \end{cases}$$

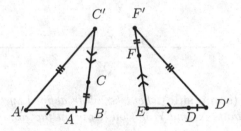

Definition 12 (in angle)

$$P \text{ in } \angle ABC := A \neq B \wedge C \neq B \wedge P \neq B \wedge$$
$$\exists X, \beta\,A\,X\,C \wedge (X = B \vee Out\,B\,X\,P)$$

Definition 13 (angle comparison)

$$\angle ABC \leq \angle DEF := \exists P, P \text{ in } \angle DEF \wedge \angle ABC \cong \angle DEP$$

Chapter 12: Parallelism. Tarski defines a strict parallelism over two pairs of points:

Definition 14 (parallelism)

$$AB \parallel CD := A \neq B \wedge C \neq D \wedge \neg\exists X, Col\,X\,A\,B \wedge Col\,X\,C\,D$$

A Proof Example. In this section we give an example of a proof. In [9], Tarski and his co-authors proves that given two angles, one is less or equal to the other one:

Theorem 1 (*lea_cases*)

$$\forall ABCDEF, A \neq B \Rightarrow C \neq B \Rightarrow D \neq E \Rightarrow F \neq E$$
$$\Rightarrow \angle ABC \leq \angle DEF \vee \angle DEF \leq \angle ABC$$

To prove the lemma *lea_cases*, Tarski implicitly uses the fact that given a line l, two points not on l are either on the same side of l or on opposite sides. But he does not give explicitly a proof of this fact. Tarski proved that if two points are on opposite sides of a line, they are not on the same side (lemma $l9\_9$), and if two points are on the same side, they are not on opposite sides (lemma $l9\_9\_bis$).

To prove that two points are either on the same side of a line, or on opposite sides, we need to show that if two points are not on opposite sides of a line they are on the same side which is the reciprocal lemma of $l9\_9\_bis$.

We will show the main steps necessary to prove that two points not on a given line l and not on opposite sides of l are on the same side:

Lemma (*not_two_sides_one_side*)

$$\neg Col\ ABX \Rightarrow \neg Col\ ABY \Rightarrow \neg A\frac{X}{Y}B \Rightarrow A\frac{}{XY}B$$

Table 2. Lemmas used in the proof

Lemma 1 (*l8_21*)

$$\forall ABC, A \neq B \Rightarrow \exists P, \exists T, AB \perp PA \wedge Col\ ABT \wedge \beta\ CTP$$

Lemma (*or_bet_out*)

$$\forall ABC, A \neq B \Rightarrow C \neq B \Rightarrow \beta\ ABC \vee Out\ BAC \vee \neg Col\ ABC$$

Lemma (*l8_18_existence*[3])

$$\forall ABC, \neg Col\ ABC \Rightarrow \exists X, Col\ ABX \wedge AB \perp CX$$

Lemma (*perp_perp_col*)

$$\forall AB\ XY\ P, P \neq A \Rightarrow Col\ ABP \Rightarrow AB \perp XP \Rightarrow PA \perp YP \Rightarrow Col\ YXP$$

Lemma (*out_one_side*)

$$\forall ABXY, (\neg Col\ ABX \vee \neg Col\ ABY) \Rightarrow Out\ AXY \Rightarrow A\frac{}{XY}B$$

Lemma (*l8_8_2*)

$$\forall PQABC, P\frac{A}{C}Q \Rightarrow P\frac{}{AB}Q \Rightarrow P\frac{B}{C}Q$$

Proof. The lemmas used in this proof are shown on Table 2. They are given without a proof.

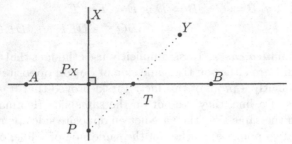

Step One
First we build the point P_X on the line AB such that $XP_X \perp AB$. The existence of P_X is proved by the lemma *l8_18_existence* (the lemmas used in this proof are provided in Table 2).

Step Two
To prove that the points X and Y are on the same side of the line AB we prove the existence of a point P verifying $A\underset{P}{\overset{X}{-\!\!\!-}}B \wedge A\underset{P}{\overset{Y}{-\!\!\!-}}B$ as required by the definition of the relation "same side" (Definition 9).

The key step of the proof is the lemma *l8_21* which allows to build such a point P. Then we will establish that this point P verifies the expected property.

To use the lemma *l8_21* we need a point on the line AB different from P_X. Since $A \neq B$, the point P_X must be different from A or B. For our proof we suppose that $P_X \neq A$. The same proof could be done using B instead of A.

Thus we can instantiate the lemma *l8_21* with the points P_X, A and Y:

$$P_X \neq A \Rightarrow \exists P, \exists T, P_X A \perp PP_X \wedge Col\, P_X\, AT \wedge \beta\, Y\, T\, P$$

Step Three
We can trivially prove that Y and P are located on opposite sides of the line AB since T is collinear with A and Px. Therefore T is collinear with A and B, and T is between P and Y which correspond exactly to the definition of the "opposite sides" relation (Definition 8). Thus we get:

$$A\underset{Y}{\overset{P}{-\!\!\!-\!\!\!-}}B \tag{1}$$

Step Four
Now it remains to show that X and P are located on opposite sides of the line AB. First, we prove that X, Px and P are collinear using the lemma *perp_perp_col*.

Second, we use the lemma *or_bet_out* applied to the three points X, P_X and P to distinguish three cases:

    1. $\beta\, X\, P_X\, P$
    2. $Out\, P_X\, X\, P$
    3. $\neg Col\, X\, P_X\, P$

1. The first case gives trivially a proof of $A\overset{X}{\underset{P}{\rule{2em}{0.4pt}}}B$ since $\beta\ X P_X\ P$ and $Col\ A B\ P_X$ which is the definition of the relation "opposite sides" (Definition 8). Since $A\overset{X}{\underset{P}{\rule{2em}{0.4pt}}}B$ and $A\overset{Y}{\underset{P}{\rule{2em}{0.4pt}}}B$ (step 3) we can conclude $A\underset{XY}{\rule{2em}{0.4pt}}B$ using the definition of the relation "same side" (Definition 9).

2. The second case also leads to a contradiction:
   The lemma *out_one_side* allows us to deduce $A\underset{PX}{\rule{2em}{0.4pt}}B$.
   Using *out_one_side* applied to $P_X\ A X\ P$ we have:

$$(\neg Col\ P_X\ A X \vee \neg Col\ P_X\ A P) \Rightarrow Out\ P_X\ X P \Rightarrow P_X\underset{XP}{\rule{2em}{0.4pt}}A$$

Since P_X is collinear with A and B we also get:

$$A\underset{XP}{\rule{2em}{0.4pt}}B \Longleftrightarrow A\underset{PX}{\rule{2em}{0.4pt}}B \quad \text{(symmetry of "same side")} \qquad (2)$$

Finally, we will derive the contradiction using lemma $l8\_8\_2$:
Using $l8\_8\_2$ applied to A, B, P, X and Y, we get:

$$A\underbrace{\overset{P}{\underset{Y}{\rule{2em}{0.4pt}}}}_{(1)}B \Rightarrow A\underbrace{\underset{PX}{\rule{2em}{0.4pt}}}_{(2)}B \Rightarrow A\overset{X}{\underset{Y}{\rule{2em}{0.4pt}}}B$$

The hypothesis $\neg A\overset{X}{\underset{Y}{\rule{2em}{0.4pt}}}B$ is in contradiction with the conclusion $A\overset{X}{\underset{Y}{\rule{2em}{0.4pt}}}B$.

3. The third case leads easily to a contradiction since we proved $Col\ X\ P_X\ P$.
   $\qquad\qquad\qquad\qquad\qquad\qquad\qquad\qquad\qquad\qquad\qquad\qquad\qquad\qquad\qquad$ □

3   Hilbert's Axiom System

Hilbert's axiom system is based on two abstract types: points and lines (as we limit ourselves to 2-dimensional geometry we do not introduce 'planes' and the related axioms). In Coq's syntax we have:

```
Point : Type
Line  : Type
```

We assume that the type Line is equipped with an equivalence relation EqL which denotes equality between lines:

```
EqL      : Line -> Line -> Prop
EqL_Equiv : Equivalence EqL
```

We do not use Leibniz equality (the built-in equality of Coq), because when we will define the notion of line inside Tarski's system, equality will be a defined notion. Note that we do not follow closely the Hilbert's presentation because we use an explicit definition of the equality relation.

We assume that we have a relation of incidence between points and lines:

```
Incid : Point -> Line -> Prop
```

We also assume that we have a relation of betweenness:

```
BetH : Point -> Point -> Point -> Prop
```

Notice that contrary to the `Bet` relation of Tarski, the one of Hilbert implies that the points are distinct.

The axioms are classified by Hilbert into five groups: Incidence, Order, Parallel, Congruence and Continuity. We formalize here only the first four groups, leaving out the continuity axiom. We provide details only when the formalization is not straightforward.

3.1   Incidence Axioms

Axiom (I 1). For every two distinct points A, B there exists a line l such that A and B are incident to l.

```
line_existence : forall A B, A<>B -> exists l, Incid A l /\ Incid B l;
```

Axiom (I 2). For every two distinct points A, B there exists at most one line l such that A and B are incident to l.

```
line_unicity : forall A B l m, A <> B ->
           Incid A l -> Incid B l -> Incid A m -> Incid B m -> EqL l m;
```

Axiom (I 3). There exists at least two points on a line. There exists at least three points that do not lie on a line.
```
two_points_on_line : forall l, exists A, exists B,
                           Incid B l /\ Incid A l /\ A <> B

ColH A B C := exists l, Incid A l /\ Incid B l /\ Incid C l

plan : exists A, exists B, exists C, ~ ColH A B C
```

3.2   Order Axioms

It is straightforward to formalize the axioms of order:

Axiom (II 1). If a point B lies between a point A and a point C then the point A,B,C are three distinct points through a line, and B also lies between C and A.

```
between_col  : forall A B C : Point, BetH A B C -> ColH A B C
between_comm : forall A B C : Point, BetH A B C -> BetH C B A
```

Axiom (II 2). For two distinct points A and B, there always exist at least one point C on line AB such that B lies between A and C.

```
between_out :  forall A B  : Point,
                   A <> B -> exists C : Point, BetH A B C
```

Axiom (II 3). Of any three distinct points situated on a straight line, there is always one and only one which lies between the other two.

```
between_only_one : forall A B C : Point,
                        BetH A B C -> ~ BetH B C A /\ ~ BetH B A C

between_one : forall A B C, A<>B -> A<>C -> B<>C -> ColH A B C ->
                        BetH A B C \/ BetH B C A \/ BetH B A C
```

Axiom (II 4 - Pasch). *Let A, B and C be three points that do not lie in a line and let a be a line (in the plane ABC) which does not meet any of the points A, B, C. If the line a passes through a point of the segment AB, it also passes through a point of the segment AC or through a point of the segment BC.*

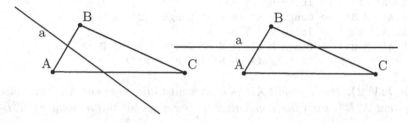

To give a formal definition for this axiom we need an extra definition:

```
cut l A B := ~Incid A l /\ ~Incid B l /\
                        exists I, Incid I l /\ BetH A I B

pasch : forall A B C l, ~ColH A B C -> ~Incid C l -> cut l A B ->
                        cut l A C \/ cut l B C
```

3.3   Parallel Axiom

As the formalization is done in a two-dimensional setting, we follow Hilbert and say that two lines are parallel when they have no point in common. Then Euclid's axiom states that there exists a unique line parallel to another line *l* passing through a given point *P*. Note that as the notion of parallel is strict we need to assume that *P* does not belong to *l*.

```
Para l m := ~ exists X, Incid X l /\ Incid X m;
euclid_existence : forall l P, ~ Incid P l -> exists m, Para l m;
euclid_unicity :  forall l P m1 m2, ~ Incid P l ->
                        Para l m1 -> Incid P m1 ->
                        Para l m2 -> Incid P m2 -> EqL m1 m2;
```

3.4   Congruence Axioms

The congruence axioms are the most difficult to formalize because Hilbert does not provide clear definitions for all the concepts occurring in the axioms. Here is the first axiom:

Axiom (IV 1). *If A, B are two points on a straight line a, and if A' is a point upon the same or another straight line a', then, upon a given side of A' on the straight line a', we can always find one and only one point B' so that the segment AB is congruent to the segment A'B'. We indicate this relation by writing AB ≡ A'B'.*

To formalize the notion of "on a given side", we split the axiom into two parts: existence and uniqueness. We state the existence of a point on each side, and we state the uniqueness of this pair of points.

```
cong_existence : forall A B l M, A <> B -> Incid M l ->
   exists A', exists B',
   Incid A' l /\ Incid B' l /\ BetH A' M B' /\
                                   CongH M A' A B /\ CongH M B' A B
```

```
cong_unicity : forall A B l M A' B' A'' B'', A <> B -> Incid M l ->
   Incid A'  l    -> Incid B'  l ->
   BetH A'  M B'  -> CongH M A'   A B -> CongH M B'   A B ->
   Incid A'' l    -> Incid B'' l ->
   BetH A'' M B'' -> CongH M A''  A B -> CongH M B''  A B ->
   (A' = A'' /\ B' = B'') \/ (A' = B'' /\ B' = A'')
```

Axiom (IV 2). *If a segment AB is congruent to the segment $A'B'$ and also to the segment $A''B''$, then the segment $A'B'$ is congruent to the segment $A''B''$.*

The formalization of this axiom is straightforward:

```
cong_pseudo_transitivity : forall A B A' B' A'' B'',
     CongH A B A' B' -> CongH A B A'' B'' -> CongH A' B' A'' B''
```

Note that from the last two axioms we can deduce the reflexivity of the relation $\equiv$.

Axiom (IV 3). *Let AB and BC be two segments of a straight line a which have no points in common aside from the point B, and, furthermore, let $A'B'$ and $B'C'$ be two segments of the same or of another straight line a' having, likewise, no point other than B' in common. Then, if $AB \equiv A'B'$ and $BC \equiv B'C'$, we have $AC \equiv A'C'$.*

First, we define when two segments have no common points. Note that we do not introduce a type of segments for the sake of simplicity.

```
Definition disjoint A B C D :=
                    ~ exists P, Between_H A P B /\ Between_H C P D.
```

Then, we can formalize the axioms IV 3:

```
addition: forall A B C A' B' C',
   ColH A B C -> ColH A' B' C' ->
   disjoint A B B C -> disjoint A' B' B' C' ->
   CongH A B A' B' -> CongH B C B' C' -> CongH A C A' C'
```

Angle. Hilbert defines an angle with two distinct half-lines emanating from a same point. The imposed condition that two half-lines be distinct excludes the null angle from the definition. Tarski defines an angle with three points. Two of them have to be different from the third which is the top of the angle. Such a definition allows null angles. For our formalization of Hilbert, we choose to differ slightly from his definition and use a triple of points. Our definition includes the null angle. Defining angles using half-lines consists in a definition involving four points and the proof that two of them are equal. It is just simpler to use only three points.

```
Record Triple {A:Type} : Type :=
    build_triple {V1  : A ;
                  V   : A ;
                  V2  : A ;
                  Pred : V1 <> V /\ V2 <> V}.
```

```
Definition angle := build_triple Point.
```

Axiom (IV-4). *Given an angle* α*, a half-line* h *emanating from a point* O *and given a point* P*, not on the line generated by* h*, there is a unique half-line* h' *emanating from* O*, such that the angle* α' *defined by* (h, O, h') *is congruent with* α *and such that every point inside* α' *and* P *are on the same side with respect to the line generated by* h*.*

To formalize this axiom we need definitions for the underlying concepts.

Hilbert uses the "same side" notion to define interior points of an angle:

> Given two half-lines h and h' emanating from a same point, every point P on the same side of h as a point of h' and on the same side of h' as a point of h is in the interior of the angle defined by h and h'.

Hilbert gives a formal definition of the relative position of three points on the same line:

$$\forall A\,O\,B\,, \beta_H(A, O, B) \iff A \text{ and } B \text{ are on opposite sides of } O \tag{3}$$

$$\forall A\,A'\,O\,,\ \beta_H(A, A', O) \vee \beta_H(A', A, O) \iff A \text{ and A' are on the same side of } O \tag{4}$$

This second definition (4) allows to define the notion of half-line: given a line l, and a point O on l, all pairs of points laying on the same side of O belong to the same half-line emanating from O.

```
outH P A B := BetH P A B \/ BetH P B A \/ (P <> A /\ A = B);
```

We define the interior of an angle as following:

```
InAngleH a P :=
 (exists M, BetH (V1 a) M (V2 a) /\
 ((outH (V a) M P) \/ M = (V a))) \/
   outH (V a) (V1 a) P \/
   outH (V a) (V2 a) P;
```

Hilbert gives a formal definition of the relative position of two points and a line:

```
same_side A B l := exists P, cut l A P /\ cut l B P;
```

Then the fourth axiom is a little bit verbose because we need to manipulate the non-degeneracy conditions for the existence of the angles and half-lines.

```
aux : forall (h h1 : Hline), P1 h = P1 h1 -> P2 h1 <> P1 h;
hcong_4_existence: forall a h P,
  ~Incid P (line_of_hline h) -> ~ BetH (V1 a)(V a)(V2 a) ->
  exists h1, (P1 h) = (P1 h1) /\ (forall CondAux : P1 h = P1 h1,
  CongaH a (angle (P2 h) (P1 h) (P2 h1)) (conj (sym_not_equal (Cond h))
  (aux h h1 CondAux)))  /\
  (forall M, ~ Incid M (line_of_hline h) /\ InAngleH (angle (P2 h) (P1 h) (P2 h1)
    (conj (sym_not_equal (Cond h)) (aux h h1 CondAux))) M ->
  same_side P M  (line_of_hline h)));
```

The uniqueness axiom requires an equality relation between half-lines[4]:

```
hEq : relation Hline := fun h1 h2 => (P1 h1) = (P1 h2) /\
        ((P2 h1) = (P2 h2) \/ BetH (P1 h1) (P2 h2) (P2 h1) \/
                         BetH (P1 h1) (P2 h1) (P2 h2));

hline_construction a (h: Hline)  P (hc:Hline) H :=
 (P1 h) = (P1 hc) /\
 CongaH a (angle (P2 h) (P1 h) (P2 hc)) (conj (sym_not_equal (Cond h)) H)) /\
   (forall M, InAngleH (angle (P2 h) (P1 h) (P2 hc)
                              (conj (sym_not_equal (Cond h)) H)) M ->
   same_side P M  (line_of_hline h));

hcong_4_unicity :  forall a h P h1 h2 HH1 HH2,
   ~Incid P (line_of_hline h) -> ~ BetH (V1 a)(V a)(V2 a) ->
   hline_construction a h P h1 HH1 -> hline_construction a h P h2 HH2 ->
   hEq h1 h2
```

The last axiom is easier to formalize as we already have all the required definitions:

Axiom (IV 5). *If the following congruences hold* $AB \equiv A'B'$, $AC \equiv A'C'$, $\angle BAC \equiv \angle B'A'C'$ *then* $\angle ABC \equiv \angle A'B'C'$

```
cong_5 : forall A B C A' B' C',
  forall H1 : (B<>A /\ C<>A),
  forall H2 : (B'<>A' /\ C'<>A'),
  forall H3 : (A<>B /\ C<>B),
  forall H4 : (A'<>B' /\ C'<>B'),
  CongH A B A' B' -> CongH A C A' C' ->
  CongaH (angle B A C H1) (angle B' A' C' H2) ->
  CongaH (angle A B C H3) (angle A' B' C' H4)
```

---

[4] P1 is the function to access to the first point of the half-line and P2 the second point.

3.5 Comparison with Phil Scott's Formalization

Giving a formal definition for these axioms is not trivial and thus error prone. Phil Scott noticed in [6] that Laura Meikle made an error during her formalization in 2003. Hence, we compared our formalization of the axioms to the one of Scott in Isabelle. Most axioms are formalized similarly. Sometimes we have a slightly stronger version: for the axiom *IV-5*, he assumes that the points A,B and C are not collinear and A',B' and C' are not collinear either. The axiom *IV-1* is formalized differently in [6]. It is difficult to know what was the intend of Hilbert, so we also proved his version using the axioms of Tarski:

```
Definition same_side_scott E A B :=
 E <> A /\ E <> B /\ Col_H E A B /\ ~ Between_H A E B.
```

```
Remark axiom_hcong_scott:
 forall P Q A C, A <> C -> P <> Q ->
   exists B, same_side_scott A B C  /\ Hcong P Q A B.
```

Our formalization differs also in the definition of InAngle for the axiom *IV-4*.

4 Hilbert Follows from Tarski

In this section, we describe the main result of our development, which consists in a formal proof that our chosen formalization of Hilbert's axioms can be defined and proved within Tarski's axiom system. We prove that Tarksi's system constitutes a model of Hilbert's axioms (continuity axioms are excluded from this study).

Encoding the Concepts of Hilbert within Tarski's Geometry. In this section, we describe how we can define the different concepts involved in Hilbert's axiom system using the definition of Tarski. We also compare the definitions in the two systems when they are not equivalent. We will define the concepts of line, betweenness, out, parallel, angle.

Lines: To define the concept of line within Tarski, we need the concept of two distinct points. For our formalization in Coq, we use a dependent type which consists in a record containing two elements of a given type A together with a proof that they are distinct. We use a polymorphic type instead of defining directly a couple of points for a technical reason. To show that we can instantiate Hilbert type class in the context of Tarski, Coq will require that some definitions in the two corresponding type classes share the definition of this record.

```
Record Couple {A:Type} : Type :=
                    build_couple {P1: A ; P2 : A ; Cond: P1 <> P2}.
```

Then, we can define a line by instantiating A with the type of the points to obtain a couple of points:

```
Definition Line := @Couple Tpoint.
Definition Lin := build_couple Tpoint.
```

But, if for example we have four distinct points A, B, C and D which are collinear, the lines AB and CD are different according to Leibniz equality (the standard equality of Coq), hence we need to define our own equality on the type of lines:

```
Definition Eq : relation Line :=
                fun l m => forall X, Incident X l <-> Incident X m.
```

We can easily show that this relation is an equivalence relation. And we also show that it is a proper morphism for the `Incident` predicate.

```
Lemma eq_incident : forall A l m,
                        Eq l m -> (Incident A l <-> Incident A m).
```

Betweenness: As noted before, Hilbert's betweenness definition differs from Tarski's one. Hilbert define a strict betweenness which requires that the three points concerned by the relation to be different. With Tarski, this constraint does not appear. Hence we have:

```
Definition Between_H A B C := Bet A B C /\ A <> B /\ B <> C /\ A <> C.
```

Out: Here is a definition of the concept of 'out' defined using the concepts of Hilbert:

```
Definition outH :=
  fun P A B => Between_H P A B \/ Between_H P B A \/ (P <> A /\ A = B).
```

We can show that it is equivalent to the concept of 'out' of Tarski:

```
 Lemma outH_out : forall P A B, outH P A B <-> out P A B.
```

Parallels: The concept of parallel lines in Tarski's formalization includes the case where the two lines are equal, whereas it is excluded in Hilbert's. Hence we have:

```
Lemma Para_Par : forall A B C D, forall HAB: A<>B, forall HCD: C<>D,
            Para (Lin A B HAB) (Lin C D HCD) -> Par A B C D
```

where **par** denotes the concept of parallel in Tarski's system and **Para** in Hilbert's. Note that the converse is not true.

Angle: As noted before we define an angle by a triple of points with some side conditions. We use a polymorphic type for the same reason as for the definition of lines:

```
Record Triple {A:Type} : Type :=
     build_triple {V1  : A ;
                   V   : A ;
                   V2  : A ;
                   Pred : V1 <> V /\ V2 <> V}.

Definition angle := build_triple Tpoint.

Definition InAngleH a P :=
  (exists M, Between_H (V1 a) M (V2 a) /\ ((outH (V a) M P) \/ M=(V a)))
                       \/ outH (V a) (V1 a) P \/ outH (V a) (V2 a) P.

Lemma in_angle_equiv : forall a P, (P <> (V a) /\ InAngleH a P) <->
                                    InAngle P (V1 a) (V a) (V2 a).
```

Main Result. Once the required concepts have been defined, we can use our large set of results describe in Sec. 2.2 to prove every axiom of Hilbert's system. To capture our result within Coq we assume we have an instance T of the class Tarski, and we show that we have an instance of the class Hilbert: This requires 1200 lines of formal proof. From a technical point, to capture this fact in Coq, we could have built a *functor* from a module type to another module type. We chose the approach based on type classes, because type classes are first class citizens in Coq.

```
Section Hilbert_to_Tarski.

Context '{T:Tarski}.

Instance Hilbert_follow_from_Tarski : Hilbert.
Proof.
 ... (* omitted here *)
Qed.

End Hilbert_to_Tarski.
```

5   Conclusion

We have proposed the formal proof that our chosen formalization of Hilbert's axioms can be derived from Tarski's axioms. This work can now serve as foundations for the many other Coq developments about geometry. The advantage of Tarski's axioms lies in the fact that there are few axioms and most of them have been shown to be independent from the others. Moreover a large part of our proofs are independent of some axioms. For instance the axiom of Euclid is used for the first time in Chapter 12. Hence, the proofs derived before this chapter are also valid in absolute geometry. In the future we plan to reconstruct the foundations of Frédérique Guilhot's formalization of high-school geometry and of our formalizations of automated deduction methods in geometry [17, 19] using Tarski's axioms. Another natural extension of this work would be to prove Tarski's axioms using Hilbert's axioms. This would be a excellent way to validate our formal interpretation of Hilbert's axioms.

Availability

The full Coq development consists of more than 500 lemmas and 23000 lines of formal proof. The formal proofs and definitions with hypertext links and dynamic figures can be found at the following urls:

http://dpt-info.u-strasbg.fr/~narboux/tarski.html
http://gabrielbraun.free.fr/Geometry/Tarski/

Acknowledgments. We would like to thank our reviewers for their numerous remarks which helped improve this paper.

References

1. Avigad, J., Dean, E., Mumma, J.: A formal system for euclid's elements. Review of Symbolic Logic 2, 700–768 (2009)
2. Coq development team: The Coq Proof Assistant Reference Manual, Version 8.3. TypiCal Project (2010)
3. Dehlinger, C., Dufourd, J.-F., Schreck, P.: Higher-Order Intuitionistic Formalization and Proofs in Hilbert's Elementary Geometry. In: Richter-Gebert, J., Wang, D. (eds.) ADG 2000. LNCS (LNAI), vol. 2061, pp. 306–324. Springer, Heidelberg (2001)
4. Paulson, L.C.: The Isabelle reference manual (2006)
5. Meikle, L.I., Fleuriot, J.D.: Formalizing Hilberts Grundlagen in Isabelle/Isar. In: Basin, D., Wolff, B. (eds.) TPHOLs 2003. LNCS, vol. 2758, pp. 319–334. Springer, Heidelberg (2003)
6. Scott, P.: Mechanising hilbert's foundations of geometry in isabelle. Master's thesis, University of Edinburgh (2008)
7. Scott, P., Fleuriot, J.: An Investigation of Hilbert's Implicit Reasoning through Proof Discovery in Idle-Time. In: Schreck, P., Narboux, J., Richter-Gebert, J. (eds.) ADG 2010. LNCS, vol. 6877, pp. 182–200. Springer, Heidelberg (2011)
8. Tarski, A., Givant, S.: Tarski's system of geometry. The Bulletin of Symbolic Logic 5(2) (June 1999)
9. Schwabhäuser, W., Szmielew, W., Tarski, A.: Metamathematische Methoden in der Geometrie. Springer (1983) (in German)
10. Quaife, A.: Automated development of Tarski's geometry. Journal of Automated Reasoning 5(1), 97–118 (1989)
11. Gupta, H.N.: Contributions to the axiomatic foundations of geometry. PhD thesis, University of California, Berkley (1965)
12. Guilhot, F.: Formalisation en Coq et visualisation d'un cours de géométrie pour le lycée. TSI 24, 1113–1138 (2005) (in French)

13. Pham, T.M.: Description formelle de propriété géométriques. PhD thesis, Université de Nice - Sophia-Antipolis (2011)
14. Chou, S.C., Gao, X.S., Zhang, J.Z.: Machine Proofs in Geometry. World Scientific (1994)
15. Narboux, J.: A decision procedure for geometry in Coq. In: Slind, K., Bunker, A., Gopalakrishnan, G.C. (eds.) TPHOLs 2004. LNCS, vol. 3223, pp. 225–240. Springer, Heidelberg (2004)
16. Narboux, J.: Formalisation et automatisation du raisonnement géométrique en Coq. PhD thesis, Université Paris Sud (September 2006) (in French)
17. Janicic, P., Narboux, J., Quaresma, P.: The Area Method: a Recapitulation. Journal of Automated Reasoning 48(4), 489–532 (2012)
18. Chou, S.C.: Mechanical Geometry Theorem Proving. D. Reidel Publishing Company (1988)
19. Génevaux, J.-D., Narboux, J., Schreck, P.: Formalization of Wu's simple method in Coq. In: Jouannaud, J.-P., Shao, Z. (eds.) CPP 2011. LNCS, vol. 7086, pp. 71–86. Springer, Heidelberg (2011)
20. Pham, T.-M., Bertot, Y., Narboux, J.: A Coq-Based Library for Interactive and Automated Theorem Proving in Plane Geometry. In: Murgante, B., Gervasi, O., Iglesias, A., Taniar, D., Apduhan, B.O. (eds.) ICCSA 2011, Part IV. LNCS, vol. 6785, pp. 368–383. Springer, Heidelberg (2011)
21. Narboux, J.: Mechanical Theorem Proving in Tarski's geometry. In: Botana, F., Recio, T. (eds.) ADG 2006. LNCS (LNAI), vol. 4869, pp. 139–156. Springer, Heidelberg (2007)
22. Sozeau, M., Oury, N.: First-class type classes. In: Mohamed, O.A., Muñoz, C., Tahar, S. (eds.) TPHOLs 2008. LNCS, vol. 5170, pp. 278–293. Springer, Heidelberg (2008)

Realizations of Volume Frameworks

Ciprian S. Borcea[1,*] and Ileana Streinu[2,**]

[1] Department of Mathematics, Rider University, Lawrenceville, NJ 08648, USA
borcea@rider.edu
[2] Department of Computer Science, Smith College, Northampton, MA 01063, USA
istreinu@smith.edu, streinu@cs.umass.edu

Abstract. A *volume framework* is a $(d+1)$-uniform hypergraph together with real numbers associated to its edges. A *realization* is a labeled point set in R^d for which the volumes of the d-dimensional simplices corresponding to the hypergraph edges have the pre-assigned values. A framework realization (shortly, a framework) is *rigid* if its underlying point set is determined locally up to affine volume-preserving transformations. If it ceases to be rigid when any volume constraint is removed, it is called *minimally rigid*.

We present a number of results on volume frameworks: a counterexample to a conjectured combinatorial characterization of minimal rigidity and a first enumerative lower bound. We also give upper bounds for the number of realizations of generic minimally rigid volume frameworks, based on degrees of naturally associated Grassmann varieties.

Keywords: volume framework, minimal rigidity, Grassmann variety.

Introduction

In this paper we study volume frameworks and their complex analogues. They are generalizations of the classical notion of bar-and-joint frameworks from Rigidity Theory. They inherit and thereby expand a central issue in rigidity theory: the characterization of all *minimally rigid structures*. In the classical case, these would be redundancy-free graphs of bars which, for generic lengths, determine locally the whole structure up to Euclidean motions; in our versions they are redundancy-free hypergraphs of simplices which, for generic volume constraints, determine locally the whole structure up to real, respectively complex affine volume-preserving transformations. In other words, these rigid structures cannot deform locally in a non-trivial way, while respecting the length or volume constraints prescribed on the edges of the graph or hypergraph.

The classical bar-and-joint case has a complete answer only in dimension two. The *Maxwell count* [8] says that such a minimally rigid structure should have no

* Research supported by a DARPA "23 Mathematical Challenges" grant.
** Research supported by NSF CCF-1016988 and a DARPA "23 Mathematical Challenges" grant.

T. Ida and J. Fleuriot (Eds.): ADG 2012, LNAI 7993, pp. 110–119, 2013.
© Springer-Verlag Berlin Heidelberg 2013

more than $2n' - 3$ bars on any subset of n' vertices, while *Laman's theorem* [7] asserts that, for n vertices, a graph with $2n-3$ bars respecting Maxwell's *sparsity condition* will be rigid, for generic edge length prescriptions. In dimension $d \geq 3$, the corresponding $(d, (d + 1)d/2)$-sparsity condition remains necessary but is no longer sufficient and the problem of a combinatorial characterization of minimally rigid graphs is open.

For volume frameworks, there are necessary sparsity conditions as well: $(d, d^2 + d - 1)$ in the real case and $(2d, 2(d^2 + d - 1))$ for its complex analogue. In the complex scenario, the marked real simplices are half the real dimension of the space $R^{2d} = C^d$ and simplices are permitted multiplicity one or two[1]. Again, sparsity alone is shown not to be sufficient for minimal rigidity.

Nevertheless, absence of specific information about the implicated minimally rigid structures does not preclude finding upper bounds for the number of their different realizations for some fixed (but otherwise generic) values of the constraints. This question was addressed, for bar-and-joint frameworks, in our earlier paper [4]. Upper bounds were obtained there from degrees of naturally associated symmetric determinantal varieties called Cayley-Menger varieties [3]. In the same vein, the upper bounds obtained here for realizations of minimally rigid volume frameworks, are derived from degrees of naturally associated Grassmann varieties.

1   Volume Frameworks

We consider $n > d$ points $p_i \in R^d$, $i = 1, ..., n$. For certain *ordered subsets* $p_I = (p_{i_0}, ..., p_{i_d})$ of $d + 1$ points, to be called *marked simplices*, we consider the *signed volume*:

$$V_I(p) = det \begin{bmatrix} 1 & 1 & ... & 1 \\ p_{i_0} & p_{i_1} & ... & p_{i_d} \end{bmatrix} = det \begin{bmatrix} p_{i_1} - p_{i_0} & ... & p_{i_d} - p_{i_0} \end{bmatrix} \quad (1)$$

The collection of all marked simplices is envisaged as a $(d + 1)$-uniform hypergraph G on n vertices. When $I = (i_0, ..., i_d)$ are the indices of a marked simplex, we have a corresponding hyperedge $I \in G$. The number of hyperedges is denoted by $|G|$. With these notational conventions, we define the function associating to a set of points the collection of volumes for the edges of G:

$$F_G : (R^d)^n \to R^{|G|}$$

$$F_G(p) = (V_I(p))_{I \in G} \quad (2)$$

---

[1] This, of course, is in keeping with the fact that, from the real point of view, there are two basic polynomial invariants of $SL(d, C)$ on d vectors: the real and the imaginary part of their complex determinant.

This polynomial function is invariant under the group Γ of *affine volume-preserving transformations*:

$$T \in \Gamma \Leftrightarrow Tx = Mx + t, \quad t \in R^d, \quad M \in SL(d, R)$$

Thus, the group Γ is described via the short exact sequence:

$$R^d \to \Gamma \to SL(d, R)$$

and has dimension $d + (d^2 - 1) = d^2 + d - 1$. Transformations $T \in \Gamma$ will also be called *trivial transformations*.

When the affine span of $p_i \in R^d, i = 1, ..., n$ is the whole space, the orbit of $p \in (R^d)^n = R^{dn}$ under Γ, that is

$$\Gamma p = \{Tp = (Tp_i)_{1 \leq i \leq n} : T \in \Gamma\} \subset R^{dn}$$

defines in a neighborhood of p a submanifold of dimension $d^2 + d - 1$, hence one needs at least $dn - (d^2 + d - 1)$ volume constraints in order to make p the *only local solution* for such constraints, up to trivial transformations. For a fixed p, let us put $\nu_I = V_I(p)$, so that p satisfies the constraints:

$$F_G(p) = (\nu_I)_{I \in G}$$

for any hypergraph G. The *weighted hypergraph* $(G, (\nu_I)_{I \in G})$ will be called a *volume framework* for dimension d, and $p \in (R^d)^n$ will be called a *realization* of that framework. Sometimes this is emphasized by using the notation $G(p)$ for the configuration of points. The set of all realizations is the fiber of F_G over $(\nu_I)_{I \in G}$:

$$F_G^{-1}((\nu_I)_{I \in G}) = F_G^{-1}(F_G(p))$$

This set-up clearly resembles the one in use for bar-and-joint frameworks [1]. The differential $dF_G(p)$ of the vector-constraint function F_G at p, conceived as a $|G| \times dn$ matrix, is called the *rigidity matrix* at p. The differential of a single constraint gives:

$$dV_I(p) \cdot u = \sum_{k=0}^{d} det \begin{bmatrix} 1 & 1 & ... \, 0 & ... \, 1 \\ p_{i_0} & p_{i_1} & ... \, u_{i_k} & ... \, p_{i_d} \end{bmatrix} \tag{3}$$

The realization p of the hypergraph G is called *infinitesimally rigid* when the rigidity matrix at p has the largest rank possible, namely $dn - (d^2 + d - 1)$.

Definition 1. *A uniform $(d+1)$-hypergraph G on n vertices will be called minimally rigid for volume frameworks in R^d when it has exactly $dn - (d^2 + d + 1)$ hyperedges and, for some constraint values $(\nu_I)_{I \in G}$, allows an infinitesimally rigid realization.*

2   Examples and Counter-Examples

We illustrate here the notion of volume framework by considering *area frame-works* in the Euclidean plane. Hypergraphs are simply referred to as graphs. Examples of minimally rigid graphs are provided by triangulated triangles. We present different (global) realizations for some configurations of this type. Then, 'counter-examples' in arbitrary dimension $d \geq 2$ will show that sparsity is not sufficient for minimal rigidity. At least one other 'vertex separation' condition must hold for $n \geq d + 2$, namely, no two vertices can be implicated in exactly the same set of marked simplices: there must be one marked simplex containing one vertex, but not the other.

2.1   Triangulated Triangles

Let p_1, p_2, p_3 be the vertices of a triangle in R^2, and let $p_4, ..., p_n$ be the remaining vertices used in some triangulation of this triangle. Since one may think in terms of a triangulated sphere (with a point 'at infinity' added to the planar comple-ment of the triangle), Euler's formula shows that our triangle is partitioned into exactly $2n - 5$ triangles. We are going to prove that generically this gives an *infinitesimally rigid area framework*, hence *the graph of the triangulation is a minimally rigid graph*.

This statement and a sketch of the argument, based on 'vertex splitting', appeared in a lecture of W. Whiteley in Oberwolfach [2]. We provide here the necessary details.

Definition 2. *Let $A - p_i$ be a vertex and let us consider a sequence of $k \geq 1$ adjacent triangles with vertex p, relabelled $Av_1v_2, Av_2v_3, ..., Av_kv_{k+1}$, where all v_i, $1 \leq i \leq k + 1$ are distinct. By vertex splitting with respect to this data, we'll mean the graph obtained by introducing one new vertex B and replacing the indicated sequence of triangles with the following list:*

$$ABv_1, \ Bv_1v_2, \ Bv_2v_3, \ ..., \ Bv_kv_{k+1}, \ ABv_{k+1}$$

See Figure 1 for a simple example.

Remark: In realizations, B may be imagined close enough to A. The argument given below will use in fact the differential of the constraint function, that is the *rigidity matrix*, for $A = B$.

Proposition 3. *The G be a graph for area frameworks and $\tilde{G}$ the graph resulting from a vertex splitting construction as described above. If G is minimally rigid, so is $\tilde{G}$.*

Proof: We start with an infinitesimally rigid realization p of G for some general values of the area constrains. Thus, the rank of the rigidity matrix $dF_G(p)$ is $|G|$.

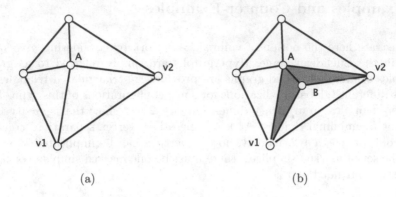

Fig. 1. A vertex splitting

The labels for vertex splitting being as above, it will be enough to prove that $dF_{\tilde{G}}(p,B)$ has rank $|\tilde{G}| = |G| + 2$ for generic $B \in R^2$. In fact, it will be enough to show that $dF_{\tilde{G}}(p,A)$ has rank $|G| + 2$.

We'll think of the rigidity matrix for $\tilde{G}$ as having the disposition of the rigidity matrix for G, with two columns added at the end for the new vertex B, and two rows added at the bottom, corresponding to triangles ABv_1 and ABv_{k+1}. Unchanged constraints determine matching rows and the row for Bv_iv_{i+1} will match the row for Av_iv_{i+1}. With V standing for signed area, we have:

$$V(Bv_iv_{i+1}) = V(Av_iv_{i+1}) + V(ABv_{i+1}) - V(ABv_i)$$

Considering the differential formula (3) for our case $A = B$, we see that adding the columns under B to the respective columns under A results in a upper-left corner identical with $dF_G(p)$ and a lower-right corner of the form

$$\begin{bmatrix} a_2 - v_{1,2} & v_{1,1} - a_1 \\ a_2 - v_{k+1,2} & v_{k+1,1} - a_1 \end{bmatrix}, \quad \text{for } A = \begin{pmatrix} a_1 \\ a_2 \end{pmatrix}, \ v_i = \begin{pmatrix} v_{i,1} \\ v_{i,2} \end{pmatrix}$$

with zero elsewhere in the last two rows. Since p is a generic realization for G, the lower-right corner is non-singular and the overall rank is $|G| + 2$.

Corollary 4. *Triangulations of a triangle define minimally rigid graphs for area frameworks.*

Indeed, a simple induction on the number of vertices shows that any triangulation of a triangle can be obtained as some sequence of vertex splitting.

2.2 Different Realizations

A planar example with two realizations is described here for six points $A, B, C,$ A', B', C'. We have seven marked triangles, namely: $ABC, A'BC, AB'C, ABC',$

$A'B'C, AB'C', A'BC'$. With triangle ABC 'pinned', vertices A', B' and C' are constrained by areas of type $*' * *$ to a line running parallel to one of the sides. The remaining constraints of type $*' *' *$ give equations of multi-degree $(1, 1, 0), (0, 1, 1)$ and $(1, 0, 1)$ on $(P_1)^3$. It follows that, in the generic case, there are two complex solutions. Thus, when one starts with a real solution, the other one will be real as well. We illustrate this in Figure 2, where the triangles ABC and $A'B'C'$ are taken to be equilateral and with the same center.

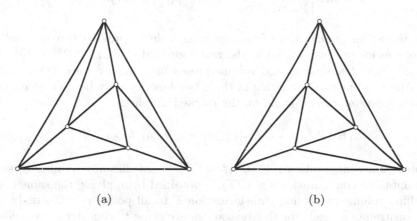

(a)   (b)

Fig. 2. Two realizations of an area framework on six vertices. The interior triangles have all been assigned area equal to one.

An iteration of this construction yields examples of area frameworks with $n = 3m$ vertices and $2^{m-1} = 2^{\frac{n}{3}-1}$ realizations.

2.3   Sparsity Is Not Enough

A graph on n vertices is (k, ℓ)-sparse when for any $n' \leq n$ vertices, the induced graph has at most $kn' - \ell$ edges. In the case of volume frameworks in R^d, a necessary condition for the linear independence of the rows in the rigidity matrix is $(d, d^2 + d - 1)$-sparsity.

However, this sparsity condition on graphs with $dn - (d^2 + d - 1)$ edges is not sufficient for minimal rigidity. One can construct counter-examples as follows. Start with a minimally rigid graph (for instance, in dimension 2, a triangulated triangle) with sufficiently many vertices. Add two more vertices and mark $2d$ new simplices, respecting sparsity but with all of them containing the two new vertices. When these two vertices slide on the same line while maintaining the distance between them, all volumes of the new simplices are preserved. Thus, the new sparse graph is not rigid.

3   Complex Volume Frameworks

We consider $n > d$ points $p_i \in C^d$, $i = 1, ..., n$. For certain ordered subsets $p_I = (p_{i_0}, ..., p_{i_d})$ of $d + 1$ points, to be called *marked simplices*, we consider the *complex volume*:

$$V_I(p) = det \begin{bmatrix} 1 & 1 & \cdots & 1 \\ p_{i_0} & p_{i_1} & \cdots & p_{i_d} \end{bmatrix} = det \begin{bmatrix} p_{i_1} - p_{i_0} & \cdots & p_{i_d} - p_{i_0} \end{bmatrix} \tag{4}$$

The *real* and *imaginary part* of this complex volume provide *two real valued* functions which are polynomial in the real coordinates of $p_{i_k} \in R^{2d} = C^d$ and *invariant* under translations and volume-preserving complex linear transformations. Any real linear combination of these two functions can be used to impose a *volume condition* or *constraint* on the marked simplex:

$$a_I Re(V_I(p)) + b_I Im(V_I(p)) = c_I, \quad a_I, b_I, c_I \in R \tag{5}$$

As noted, if a configuration $p \in (C^d)^n = (R^{2d})^n$ satisfies any number of such constraints, the configuration $Tp = (Tp_i)_i$ obtained by applying the same complex affine volume-preserving transformation T to all points p_i will satisfy the given constraints as well. For this reason, we are going to consider as *equivalent* two configurations obtained in this manner and refer to complex affine volume-preserving transformations simply as *trivial transformations*.

We denote this group of trivial transformations by Γ_C, and recall the short exact sequence

$$C^d \to \Gamma_C \to SL(d, C)$$

which gives its complex dimension as: $dim_C(\Gamma_C) = d + (d^2 - 1)$. The real dimension is therefore:

$$dim_R(\Gamma_C) = 2(d^2 + d - 1)$$

The *topological quotient* $(C^d)^n / \Gamma_C$ is not apt to carry the structure of a complex algebraic variety, but a 'categorical quotient' $(C^d)^n // \Gamma_C$ exists and gives a complex algebraic variety of complex dimension $dn - (d^2 + d - 1)$, defined up to birational equivalence [10,5,9]. We shall consider a particular birational model for $(C^d)^n // \Gamma_C$ in our discussion of upper bounds for the number of realizations.

We retain here the fact that in order to obtain isolated points in the orbit space, one has to impose *at least* $2dn - 2(d^2 + d - 1)$ *real constraints*. Of course, this necessary condition is also apparent when reasoning on the 'big' configuration space $(C^d)^n = R^{2dn}$: for a generic configuration $p \in C^{dn}$, its trivial transforms $Tp, T \in \Gamma_C$ define near p a submanifold of real codimension $2dn - 2(d^2 + d - 1)$. When as many constraints of the form (5) are imposed on configurations, with marked simplices recorded as a hypergraph G (with edge multiplicity one or two allowing for the possibility of imposing two independent real constraints on some marked simplices), we have a polynomial function:

$$F_G : R^{2dn} \to R^{2dn-2(d^2+d-1)}$$

$$F_G(p) = (a_I Re(V_I(p)) + b_I Im(V_I(p)))_{I \in G} \qquad (6)$$

If the differential $dF_G(p)$ is of *maximal rank*, that is $2dn - 2(d^2 + d - 1)$, the fiber $F_G^{-1}(F_G(p))$ of F_G through p must coincide locally with the submanifold of trivial transforms of p and one obtains the desired isolation of p among configurations with prescribed volumes $(c_I)_{I \in G} = F_G(p)$ modulo Γ_C. Under these conditions, p is called an *infinitesimally rigid* realization of the hypergraph G for prescribed constraints $(a_I, b_I, c_I)_{I \in G}$.

Definition 5. *A uniform $(d+1)$-hypergraph G on n vertices will be called minimally rigid for complex volume frameworks in $C^d = R^{2d}$ when it has exactly $2dn - 2(d^2 + d + 1)$ hyperedges and, for some constraints $(a_I, b_I, c_I)_{I \in G}$, allows an infinitesimally rigid realization.*

Obviously, this type of minimally rigid frameworks satisfy a necessary sparsity condition expressing the fact that induced hypergraphs on fewer vertices have no overconstraint.

4   Grassmann Varieties and Upper Bounds

Volume Frameworks in R^d. We consider first the case of a minimally rigid volume framework with $n \geq d + 1$ vertices and generic values for the volume constraints. A 'blunt' upper bound on the number of realizations can be obtained directly from the Bézout theorem applied as follows. Pin down one of the marked simplices with the prescribed (non-zero) volume. The remaining $n - (d+1)$ points in $R^d \subset P_d$ are determined by $dn - d(d+1)$ equations of degree at most d, hence the bound $B_{d,n} = d^{d(n-d-1)}$.

We'll obtain a more refined upper bound by observing that the determinants of all $(d + 1) \times (d + 1)$ minors in the matrix:

$$\begin{bmatrix} 1 & 1 & \dots & 1 \\ p_1 & p_2 & \dots & p_n \end{bmatrix} \qquad (7)$$

define the Plücker coordinates of the $(d + 1)$ vector subspace spanned by its rows in R^n. This point of the Grassmannian $G(d+1, n) \subset P_{\binom{n}{d+1}-1}$ contains the vector $(1, 1, ..., 1)$ and belongs therefore to a *projective cone* in $P_{\binom{n-1}{d}}$ over the Grassmannian $G(d, n-1)$. Indeed, over the complex field, this cone has dimension $d(n - d - 1) + 1$ and offers a birational model for the quotient $(C^d)^n // \Gamma_C$.

Since each constraint becomes a hyperplane section in Plücker coordinates, we obtain the bound:

$$b_{d,n} = deg\ G(d, n-1) = [d(n-d-1)]! \frac{1!2!...(n-d-2)!}{d!(d+1)!...(n-2)!}$$

from the known degree of the Grassmannian [6].

The fact that $b_{d,n}$ is a better bound than $B_{d,n}$ is easily verified by induction on $n \geq d+2$:

$$b_{d,d+2} = 1 < B_{d,d+2} = d^2$$

$$b_{d,n+1} = (dn - d^2)(dn - d^2 - 1)...(dn - d^2 - d + 1) \frac{(n-d-1)!}{(n-1)!} b_{d,n} <$$

$$< \frac{(dn - d^2)(dn - d^2 - 1)...(dn - d^2 - d + 1)}{d^d(n-1)(n-2)...(n-d)} B_{d,n+1} < B_{d,n+1}$$

We have thus proven:

Theorem 6. *For a generic choice of volume constraints, the number of realizations in R^d of a minimally rigid volume framework on n vertices is bounded from above by*

$$b_{d,n} = [d(n-d-1)]! \frac{1!2!...(n-d-2)!}{d!(d+1)!...(n-2)!} < d^{d(n-d-1)}$$

Remark: For area frameworks our bound takes the form:

$$b_{2,n} = \frac{1}{n-2} \binom{2n-6}{n-3}$$

Complex Volume Frameworks in $C^d = R^{2d}$. For minimally rigid complex volume frameworks we adapt the approach used in the real case as follows. We use a second set of complex coordinates $q \in C^d$ and record realizations not simply as $p \in C^d$, but as the conjugate pair $(p, q) = (p, \bar{p}) \in (C^d)^2$. This allows the expression of real and imaginary parts of volumes in terms of polynomials in (p, q). The quotient space C^{dn}/Γ_C is mapped, as above, by all $(d+1) \times (d+1)$ minors and their conjugates to the real points of the product:

$$\tilde{G}(d, n-1) \times \tilde{G}(d, n-1) \subset (P_{\binom{n-1}{d}})^2$$

where $\tilde{G}(d, n-1)$ denotes the projective cone over the Grassmannian $G(d, n-1)$ encountered in the real scenario. The real structure on the product is $(z, w) \mapsto (\bar{w}, \bar{z})$.

In this manner, volume constraints (5) can be interpreted as intersections with hypersurfaces of bi-degree $(1,1)$ and the number of complex solutions offers an upper bound for realizations. We obtain:

Theorem 7. *For a generic choice of volume constraints, the number of realizations in $C^d = R^{2d}$ of a minimally rigid complex volume framework on n vertices is bounded from above by*

$$\beta_{d,n} = \binom{2d(n-d-1)}{d(n-d-1)} b_{d,n}^2$$

where $b_{d,n}$ is the bound obtained above in the real case

$$b_{d,n} = [d(n-d-1)]! \frac{1!2!...(n-d-2)!}{d!(d+1)!...(n-2)!} < d^{d(n-d-1)}$$

References

1. Asimow, L., Roth, B.: The rigidity of graphs. Transactions of the American Mathematical Society 245, 279–289 (1978)
2. Bobenko, A.I., Schröder, P., Sullivan, J.M., Ziegler, G.M.: Discrete Differential Geometry. Oberwolfach Seminars, vol. 38. Birkhäuser, Basel (2008)
3. Borcea, C.S.: Point configurations and Cayley-Menger varieties. Preprint ArXiv math/0207110 (2002)
4. Borcea, C.S., Streinu, I.: The number of embeddings of minimally rigid graphs. Discrete and Computational Geometry 31, 287–303 (2004)
5. Dolgachev, I.: Lectures on Invariant Theory. London Math. Society Lecture Notes Series, vol. 296. Cambridge University Press (2003)
6. Harris, J.: Algebraic Geometry. Graduate Texts in Mathematics, vol. 133. Springer, New York (1993)
7. Laman, G.: On graphs and rigidity of plane skeletal structures. Journal of Engineering Mathematics 4, 331–340 (1970)
8. Maxwell, J.C.: On reciprocal figures, frames and diagrams of forces. Transactions of the Royal Society Edinburgh 26, 1–40 (1870)
9. Mukai, S.: An Introduction to Invariants and Moduli. Cambridege Studies in Advanced Mathematics, vol. 81. Cambridge University Press (2003)
10. Mumford, D., Fogarty, J., Kirwan, F.: Geometric Invariant Theory, 3rd edn. Springer (1994)
11. Weyl, H.: The classical groups: their invariants and representations. Princeton Landmarks in Mathematics. Princeton University Press (1997)

Rigidity of Origami Universal Molecules

John C. Bowers[1],* and Ileana Streinu[2],**

[1] Department of Computer Science, University of Massachusetts,
Amherst, MA 01003, USA
`jbowers@cs.umass.edu`
[2] Department of Computer Science, Smith College, Northampton, MA 01063, USA
`istreinu@smith.edu, streinu@cs.umass.edu`

Abstract. In a seminal paper from 1996 that marks the beginning of computational origami, R. Lang introduced TreeMaker, a method for designing origami crease patterns with an underlying metric tree structure. In this paper we address the foldability of paneled origamis produced by Lang's Universal Molecule algorithm, a key component of TreeMaker.

We identify a combinatorial condition guaranteeing rigidity, resp. stability of the two extremal states relevant to Lang's method: the initial flat, open state, resp. the folded origami base computed by Lang's algorithm. The proofs are based on a new technique of transporting rigidity and flexibility along the edges of a paneled surface.

1   Introduction

Origami, the ancient art of paper folding, is the source of challenging mathematical and computational questions. Origami *design* starts with a piece of paper, usually a convex polygon such as a square or a rectangle, on which a set of line segments (the *crease pattern*) is first drawn. The open-flat piece of paper is then folded at the creases to produce a crude three-dimensional shape, called a *base* (see Fig. 1). The base is intended to capture the general structure of some more intricate and artistic origami design, into which it will be further bent, creased and folded.

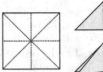

Fig. 1. A simple crease pattern on a square sheet of paper (the flat, open state) and two possible folded, projectable bases

In this paper we consider *flat-faced origami*[1], where the faces of the creased "paper" remain flat (do not bend) during folding. We may also refer to it as *paneled origami*, since it behaves like a mechanical panel-and-hinge structure. More specific definitions and preliminary concepts needed to follow our results will be given in Sec. 4.

---

* Research supported by an NSF Graduate Fellowship.
** Research supported by NSF CCF-1016988 and DARPA "23 Mathematical Challenges", under "Algorithmic Origami and Biology".
[1] Sometimes called *rigid origami* in the literature. We avoid this terminology because of its potential for ambiguity in the context of this paper.

We address one of the major open questions in algorithmic origami: when are such paneled origamis "continuously foldable"? Neither an efficient algorithm nor a good characterization are known for this decision problem, with the exception of single-vertex origami [17,16] and some disparate cases, including the "extreme bases" of [7]. In this paper, we focus on the patterns produced by Lang's *Universal Molecule* algorithm [12] for origami design.

Computational Origami. To survey all aspects of the growing literature on computational origami and its applications is beyond the scope of this paper; we will mention only the problems and directions that are related to Lang's approach. The reader may consult accessible historical surveys and articles in recent origami conference proceedings [15,18] and richly illustrated books such as [14,9]. In a seminal paper [12] that marks the beginnings of computational origami, Robert Lang formalized fundamental origami concepts, most prominently what he calls a *uniaxial base*: a flat-folded piece of paper with an underlying metric tree structure, whose faces project to a realization of the tree, called the *shadow tree*. "Flattening" the tree (i.e. aligning its edges along a single line) brings all the faces into one plane and the boundary of the original paper into a single line (the *axis* of the *uniaxial base*). Fig. 1 shows two such uniaxial bases, slightly perturbed from the flattened state to illustrate their folding patterns. Lang also described an algorithm, implemented in his freely-available TreeMaker program [13], for decomposing the paper into smaller regions called *molecules* and for computing in each of them a certain crease pattern (which Lang calls a *universal molecule*). A folded state of the complete crease pattern (obtained by putting together the universal molecules) is shown to be a uniaxial base.

Subsequent to Lang's publication of his method in 1996, several related questions emerged, aiming at solving variations of the origami design problem: NP-hardness results for general crease pattern foldability [1], the fold-and-one-cut problem [8], the disk-packing method of [10], further extensions to piecewise linear surfaces [2], etc.

TreeMaker (available at http://www.langorigami.com/) has been used for over 15 years by Lang and fellow origami artists to produce bases for complex and beautiful designs. It has inspired a wealth of new developments in mathematical and computational origami. Yet, fundamental properties of the algorithm have only appeared in sketchy form (in [12,14,9]) and a comprehensive analysis of the algorithm (both correctness and complexity) is long due. To the best of our knowledge, many of its properties still await formal proofs, and some are yet to be discovered.

Origami Design with Lang's TreeMaker. We turn now to the specific crease patterns produced by Lang's method and give it a very informal, high-level description; details will follow in Sec. 4. The input is a *metric tree*: a tree with (positive) length information attached to its edges. In Fig. 2(a), the edge lengths are calculated directly from a geometric realization (drawing) of the tree. TreeMaker is comprised of two phases: the first one solves a non-linear optimization problem, which results in a decomposition of (a minimally scaled version of) the original piece of paper into polygonal pieces, called molecules. The molecules

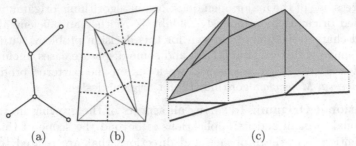

(a)   (b)   (c)

Fig. 2. The elements of Lang's Universal Molecule algorithm. (a) The input metric tree. (b) The universal molecule: a crease pattern of an input doubling polygon of the tree. (c) A folded state projecting down (along dotted lines) onto the metric tree.

have specific relationships to subtrees of the given metric tree, and, *when they are convex,* are further subdivided in the second phase of Lang's algorithm. This phase, referred to as the Universal Molecule algorithm, deals independently with each convex piece, by finding a crease pattern (called a Universal Molecule, and illustrated in Fig. 2(b)) that has a folded state (configuration) as a certain type of flat origami called a *Lang base.* The Lang base has an underlying tree structure: groups of faces (called flaps) are folded flat (overlapping) and project to a single tree edge. Fig. 2(c) shows such a Lang base state, with its *flaps* moved away from the trunk to illustrate the underlying tree structure. Lang then argues that the folded universal molecule pieces can be glued together along their boundaries to match the connectivity of the original piece of paper, but leaves open if (and when) this can be done in a non-self-intersecting manner, or when the folding of the base can be carried out continuously, without stretching, cutting, tearing or even bending of the paper.

Our Results. We present contributions to the long-due theoretical analysis of Lang's algorithm and address several open questions concerning properties of the origami crease patterns and Lang bases that it computes. We show that only certain special crease patterns *have a chance to be foldable* and identify a combinatorial pattern of a Universal Molecule (captured in an associated outerplanar graph) that forces it to be rigid in the open state and stable (not *unfoldable*) in the Lang base state. This unexpected behavior of the algorithm puts in perspective some of the most relevant properties of the computed output, and opens the way to design methods that *may* overcome these limitations. Our proof technique, called *rigidity transport,* is algorithmic in nature. As is the case with similar questions in combinatorial rigidity theory, a *complete characterization* of rigid, resp. stable patterns appears to be substantially more difficult; we leave it as an open question.

We also completed a comprehensive correctness proof of Lang's Universal Molecule algorithm by identifying and proving several useful relationships between the structures it works on, and their computational invariants. A summary of our approach to describing Lang's algorithm, which makes the identification of these invariants natural and streamlines the correctness proof, is given in Sec. 4.

Supplement. Rigidity properties of flat-faced origamis are hard to grasp from static images, without experimentation with physical or simulated models. The reader is referred to our web site http://linkage.cs.umass.edu/origamiLang, where an applet for designing Lang crease patterns and Lang bases, and our recent video [3], are available.

2   Preliminaries

Before presenting an overview of our results, we introduce some basic terminology that will be used throughout the paper.

Metric Trees, Doubling Cycles, and Doubling Polygons. A metric tree (T, w) is a tree with a weight $w(e) > 0$ assigned to each edge e. For our purposes, it is convenient to classify the tree nodes into two sets $A \cup B$, with $A = \{a_1, \cdots, a_n\}$ being the leaf nodes and $B = \{b_1, \cdots, b_m\}$ the internal ones. We also assume that a topological embedding of T is given via circular orderings of the neighbors of each internal node of the tree. A *realization* of a metric tree is given by mapping its nodes to 2D points and its edges to 2D straight-line segments, such that the length of each line segment is equal to the weight of the corresponding tree edge. The *standard realization* places each node along the x-axis and is defined as follows: root the tree T at the first leaf node a_1. Place a_1 at the origin. Then, visit each node i in breadth-first search (bfs) order from a_1. Let j be the parent of i (in the bfs) and x be the position of j. The standard realization places i at $x + w((j, i))$.

We define the *(metric) doubling cycle* C_T of T by walking around the topologically embedded tree and listing the vertices (with repetitions) in the order in which they are encountered. The vertices of the doubling cycle retain, as (not unique) labels, the labels of the tree nodes from which they came. Each vertex of a doubling-cycle thus has a *corresponding node* in the tree, a s given by this label. Each edge of the tree is traversed twice, in both directions, and appears as such in the doubling cycle. It will be assigned the length inherited from the tree; this is illustrated in Fig. 5(left). A metric realization of a doubling cycle C_T associated to a metric tree T as a planar convex polygon P_T will be called a *doubling polygon* for the tree T.

Outerplanar Graphs. These are planar graphs which can be drawn in the plane such that all of their vertices are incident to the outer face and no two edges cross. An outerplanar graph consists of a cycle $1, \cdots, n$ and other edges that, combinatorially, do not cross; we say that two edges (a, b) and (c, d) cross if c is (cyclically) between a and b (i.e. as we follow the cycle, we encounter the indices a, c, b in this order) and d is (cyclically) between b and a (or vice versa, with d between a and b and c between b and a). Such a graph has a canonical face set which is given by any planar embedding of the graph in which all vertices are incident to the outer face: we refer to this simply as *the face set* for the outer planar graph. A face which is not incident to the outer face will be referred to as being *internal*. Outerplanar graphs are used in Sec. 4 as auxiliary structures for defining a particular class of crease patterns produced by Lang's algorithm.

Origami Patterns, Realizations, and Bases. Let us consider a convex polygon P and denote by R its interior region. An *origami crease pattern* for (P, R) is a planar subdivision $G = (V, E, F)$ of $P \cup R$ into vertices V, edges E, and faces F. Each vertex $v \in V$ has a 2D coordinate, each edge e is the straight-line segment between its two vertices and has a well-defined length, and each face is a simple planar polygonal region bounded by its incident edges. An edge is incident to one or two faces; those incident to only one face are called *boundary edges* and those incident to two are *internal edges*. The union of the boundary edges is equal to the polygon P.

Let G be a crease pattern. A (3D) *realization* of the crease pattern is a mapping B of the vertices, edges, and faces of G into points, straight-line segments, and planar polygons in $\mathbf{R}^3$ such that the length of each edge is preserved and each face is mapped to a congruent planar polygon in $\mathbf{R}^3$. Intuitively, a realization is a "folded" state of the origami where each face is kept flat and folds are allowed only along the "crease" edges of the pattern. In keeping with the origami literature, we use the informal term *(origami) base* for a realization meeting some desired set of criteria.

Tree-Projectable and Lang Bases. If all the faces of a realization of an origami pattern are perpendicular to a common plane (for simplicity and without loss of generality we assume this to be the xy-plane), then the projection of the realization onto the plane is a graph. When this graph is a tree, the 3D origami realization is called a *tree-projectable base*, and the tree onto which the base projects is called the *shadow tree*. In this paper we work only with tree-projectable bases. The set of faces projecting to the same edge of the tree is called a *flap* of the base. A flap is attached to the rest of the base along perpendicular edges called *hinges*. These hinges project onto internal nodes of the tree. Flaps can be manipulated independently of the rest of the base by rotations around hinges. Flap rotations correspond to rotations of tree edges around internal nodes and vice versa. The two bases illustrated in Fig. 1(right) are tree-projectable and either is obtained from the other by simple flap rotations.

If a tree-projectable base is such that its boundary coincides with the shadow tree and it lies in the upper half plane (as in Fig. 1), then the boundary of the base can be "aligned" along a single axis by rotating the flaps until the shadow tree is in the standard alignment (as defined above). For this reason, Lang called such bases *uniaxial*. Bases computed using Lang's Universal Molecule Algorithm described in Sec. 4 satisfy these properties, but there are other crease patterns satisfying the properties which are not produced by the algorithm. To distinguish what Lang's algorithm produces from what he terms "uniaxial", we use the term *Lang base* for the output of his algorithm. Defining Lang bases independently of the algorithm would take us too far from the scope of this paper.

Configuration Space, Flexibility, and Rigidity. Given a realization of an origami pattern, new "trivial" realizations may be obtained by applying Euclidean transformations (translations and rotations) to the entire realization, but the overall "shape" of the realization remains unchanged. Factoring out translations and rotations gives us *configurations* or *origami states*. The space of all possible states (or configurations) is called the *configuration space*; formally, this is the space of all realizations modulo Euclidean translations and rotations. Paths in configuration space correspond to continuous motions of the origami pattern which keep the faces as rigid panels, which can only rotate around the edges (acting as hinges). A path in configuration space is also called a *flex*. An origami pattern in a given state is *flexible* if there exists a flex of the pattern, otherwise it is *rigid*. Rigid states are isolated points in the configuration space–given any other state, no path exists between that state and any rigid state. We use these terms to define the following concepts which will be used throughout the paper.

Flat and Open-Flat States. If all faces of an origami configuration (informally, a base) are coplanar, then we say that the base is *flat*. In this case, the dihedral angle of each internal edge is either 0 or π. The converse is also true. If further, the dihedral angle at each internal edge is π, we say that the base is in the *open-flat* state.

With these preliminary concepts in place, we turn now to a high level description of our results.

3   Overview of the Main Results

Lang's algorithm produces origami patterns that *may fold* into Lang bases, i.e. which have a state that is "uniaxial". We will show that many of these patterns are in fact *rigid:* they are isolated points in the origami's configuration space, and therefore do not fold to anything else. The corresponding (folded) Lang base lies in a different component of the configuration space, and, due to its intrinsic tree-like structure, it is obviously flexible. However, we will show that it is also isolated, but in a different way, which we'll call *stable*.

To prove these results, we had to complete a theoretical analysis, including correctness, of Lang's Universal Molecule algorithm. Lang's ideas are beautiful and sound, but other than a very sketchy sequence of statements, no comprehensive proof was published since the algorithm was first announced in 1996. Since the first phase of TreeMaker may sometimes fail, there was no guarantee (other than perhaps statistical evidence from running the code) that the second phase (the Universal Molecule) would not come across some special situation where it would also fail. We will therefore rely, throughout this paper, on the following result; its statement requires additional concepts, which will be defined later in Section 4.

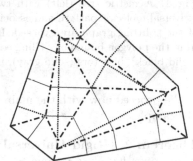

Fig. 3. Rigid crease pattern. The mountain (dot-dashed) and valley (dotted) assignment indicates the pattern of the flat-folded Lang base configuration.

Theorem 1. (Correctness of Lang's Universal Molecule Algorithm [12,4]) *Given an input metric tree and a Lang polygon associated to it, Lang's universal molecule algorithm correctly computes a crease pattern which has a second, flat-folded realization as a Lang base which projects exactly to the input metric tree.*

The proof required us to clarify the role of various structures appearing in the algorithm (in particular, what we call a Lang polygon associated to a metric tree) and to identify and prove relationships (invariants) between them. For instance, we need the property that the *splitting edges*, introduced by Lang's algorithm to reconcile the metric properties of the input tree within the constraints of the polygonal piece of paper, *do not cross*. This permits the algorithm to proceed recursively in an un-equivocal manner. We clarify the role of the perpendicular creases and track them during the algorithm's execution. By contrast, Lang defines them at a post-computation step, when they are extended recursively in a manner that was never shown to let them "arrive" precisely at specific points, at a well-defined distance from the polygon corners. A proof sketch is found in Sec. 4, and details are given in [4].

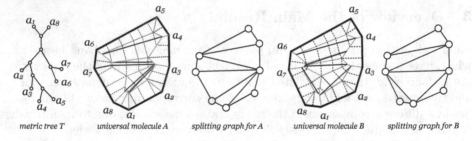

Fig. 4. A metric tree (left) with two compatible Lang polygons, each one with its universal molecule pattern and associated splitting graph. In the first case, all the faces of the splitting graph are exposed. In the second case, one triangular face is isolated from the polygon boundary. Tiling edge decorations indicate edge types: splitting edges (solid black), bisectors (solid gray), perpendiculars (dashed black).

We proceed now with the main results, starting with the base case of a family of examples:

Theorem 2. (Rigid Universal Molecules) *There exist universal molecules whose open-flat state is rigid.*

The existence is based on the universal molecule in Fig. 3. It is easy to verify that this example is a universal molecule by following the description of Lang's algorithm given in Sec. 4. The coordinates are *generic*, i.e. the pattern does not change under small perturbations of the defining tree and polygonal region in which the origami pattern is computed. The proof, given in the next section, is also indicative of the fact that the pattern's rigidity is not a simple artifact of some rare occurrence or numerical imprecision.

We generalize this example in two ways, first by turning it into a sufficient criterion for detecting the rigidity of the crease pattern, then by extending it to the Lang base state produced by Lang's algorithm. The generalization leads to a family of universal molecules distinguished by the existence of a special degree-6 vertex which we call an *isolated peak* (defined in Sec. 6). Informally, an isolated peak is a degree-6 vertex of the crease pattern which is "isolated" from the boundary in the manner of the basic example from Fig. 3: none of the creases emanating from the vertex reach the boundary of the crease pattern. The crease pattern shown in Fig. 5(right) has two degree-6 vertices, but neither are isolated. The meaning of "generic" in the following theorem will be made precise in Section 5.

Theorem 3. (Universal Molecules with isolated peaks are rigid) *If a generic crease pattern produced by Lang's Universal Molecule algorithm has an isolated peak, then it is rigid in the open-flat state.*

A Lang base produced by Lang's algorithm is always flexible, inheriting the same degrees of freedom as its shadow tree. A foldable base should be reached through a continuous deformation path from the open origami state. Of course, we know that Universal Molecule crease patterns with isolated peaks lead to Lang bases that cannot be reached from the open state, but could they be reached from some other interesting intermediate configuration that is not just a reconfiguration of its flaps? For instance, is it possible to separate the overlapping faces forming a flap? We prove that this is not the case.

Theorem 4. (Stability of Lang Lang Bases with isolated peaks) *If a generic crease pattern produced by Lang's Universal Molecule algorithm has an isolated peak then the Lang base associated to it cannot be unfolded, in the sense that overlapping panels grouped into flaps cannot be separated.*

4   Computing the Universal Molecule

We present now a summary of our formulation of Lang's Universal Molecule algorithm, from [4]. We start by defining the algorithm's *input* (a metric tree and a convex polygonal region, forming what we call a Lang polygon), *output* (an extended universal molecule tiling and a specific Lang base compatible with it), then describe the algorithm (as a sweep process with two types of

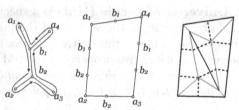

Fig. 5. A metric tree surrounded by a doubling cycle, a possible Lang molecule associated to it, and its universal molecule crease pattern

events, and with splitting and merging subroutines in a divide-and-conquer fashion at certain events), and conclude by stating the relationships maintained throughout the algorithm's execution between the input and output structures ("invariants").

Lang Polygon. A *Lang polygon* associated to a metric tree (T, w) is a convex polygonal region R_T with polygon boundary P_T satisfying the following two conditions: (a) P_T is a doubling polygon for T and (b) the distance, inside the polygonal region R_T, between two vertices corresponding to leaves a_i and a_j in the tree T is at least the distance in the tree between these leaves. These conditions imply that the vertices of the Lang polygon P_T labeled with internal nodes $b_i \in B$ of T must be *straight*, i.e. incident to an angle of π. See Fig. 5. For this reason, we refer to the vertices of P_T corresponding to leaf nodes of T as *corners* and to the internal nodes of T as *markers*, since they are simply "marked" along a line segment between corners.

Remarks. In TreeMaker, the first ("optimization") phase produces a decomposition into such polygonal regions, called here *Lang polygons* in recognition of this fact. We remark that condition (b) does not hold for all convex realizations of the doubling polygon P_T. An important invariant in our proof of correctness is that contours (defined below) remain Lang polygons. We also note that Lang does not retain markers for the inner nodes of the tree on the polygon boundary.

Parallel Sweep and Contours. A *parallel sweep* is a process in which the polygon edges are moved, parallel with themselves and at constant speed, towards the interior, as in Fig. 6. It has been used in the literature to define the *straight-skeleton* of a polygon. The process is parametrized by h, representing a *height*, or distance between a polygon edge and a line parallel to it. The moving lines cross at vertices $a_i(h)$, $i = 1, \cdots, n$, which trace the inner bisectors s_i of the polygon vertices, starting at height $h = 0$ with $a_i(0) = a_i$. This gives a continuous family of polygons $P(h)$ (called *h-contours* or simply *contours* of P), parametrized by the height h.

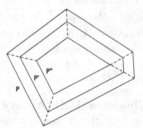

Fig. 6. Contours and an edge contraction event in a parallel sweep

A **Universal Molecule (UM)** is a specific crease pattern on a Lang polygon, made by tracing all the vertices of the polygon (corresponding to *both leaf and internal nodes* of the metric tree) during a parallel sweep with *events* described below. The distinctive property of the UM (which of course needs a proof) is that *it has an isometric realization as a Lang base whose shadow tree is isometric to the given metric tree of the Lang polygon*. Fig. 2 illustrates this correspondence.

Remark. The specific examples of universal molecule crease patterns we study in Theorems 2, 3, and 4 are created without edge contraction events (described below) and without simultaneous events: such a pattern is called *generic*. In general, edge contraction events and simultaneous events occur only for a very small subset of possible inputs to Lang's algorithm. Note that in the special (and rare) case when only edge contraction events occur, then the main structure underlying the universal molecule is simply the straight-skeleton of the molecule's convex polygon.

Edge Contraction Event. As h increases, the edges shrink until one of them reaches zero length. We call this an *(edge) contraction event*; it happens for a pair of consecutive bisectors (s_k, s_{k+1}) whose intersection c_k is at minimum distance h from the corresponding edge $a_k a_{k+1}$ of the original polygon, as in Fig. 6. Note that several events may happen simultaneously, either at the same crossing point c_k or at different ones.

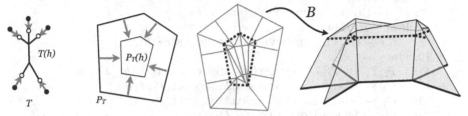

(a) Leaf nodes move inwards on their edges. Internal nodes stay fixed.

(b) The counterpart on the Lang base is a line sweep, parallel to the axis. Illustrated here is a splitting event.

Fig. 7. Parallel sweeping a Lang polygon: the effect on the metric tree, the tiling and the Lang base

Tree and Contour Polygon Shrinking. Perpendiculars dropped from a vertex of the contour polygon to the two adjacent polygon sides (on whose bisector it moves) cut off equal length segments from these sides. Interpreting this in terms of the metric tree and its doubling polygon, we remark that (a) the parallel sweep is reflected by a simultaneous constant-speed decrease in the lengths of the leaf edges in T; for each leaf, the constant depends on the angle at the corresponding Lang polygon vertex; the result is a shrunken-tree process $T(h)$; (b) the contour $P(h)$ is a (parallel) realization of the metric doubling polygon of $T(h)$. See Fig. 7(a). The distance between two leaves in the (original) tree equals the sum of the pieces removed from their corresponding leaf edges, plus the distance in the shrunken tree. The standard alignment of $T(h)$ is related to that of T by fixing the internal nodes and shrinking the leaves Fig. 7(b) shows the effect of the sweep on the Lang base and the correspondence with the shrunken tree.

Splitting Events. For the shrunken tree $T(h)$, the h-contour $P(h)$ may stop being a Lang polygon: this happens exactly when the distance between two non-consecutive corner vertices of a contour becomes equal to their distance in the shrunken h-tree $T(h)$. We call this a *splitting event*, as it splits the current Lang polygon, by a diagonal called a *splitting edge*, into two smaller Lang polygons. See Fig. 7(a).

We have now all the ingredients to present the algorithm.

Lang's Universal Molecule Algorithm. The **input** is a metric tree T and a Lang polygon R_T with boundary $P = P_T$ associated to it. The **output** is a crease pattern such that the resulting origami has a realization (called a Lang base) as a 3D state whose shadow is a standard alignment of the input metric

---

Algorithm 1. CALCULATEUNIVERSALMOLECULE(P_T, T)

1: **if** ISBASECASE(P_T, T) **then**
2:     **return** HANDLEBASECASE(P_T, T)
3: **end if**
4: $h_1 \longleftarrow$ GETNEXTCONTRACTIONEVENT(P_T)
5: $h_2 \longleftarrow$ GETNEXTBRANCHINGEVENT(P_T, T)
6: $h \longleftarrow \min\{h_1, h_2\}$
7: $G' \longleftarrow$ COMPUTEANNULUSTILING($P_T, P_T(h)$)
8: **if** $h_1 \leq h_2$ **then**
9:     $G \longleftarrow$ CALCULATEUNIVERSALMOLECULE($P_T(h), T(h)$)
10: **else**
11:     $P_1, T_1, P_2, T_2 \longleftarrow$ SPLITATBRANCHINGEVENT($P_T(h), T(h)$)
12:     $G_1 \longleftarrow$ CALCULATEUNIVERSALMOLECULE(P_1, T_1)
13:     $G_2 \longleftarrow$ CALCULATEUNIVERSALMOLECULE(P_2, T_2)
14:     $G \longleftarrow$ MERGEMOLECULES(G_1, G_2)
15: **end if**
16: **return** MERGEANNULUS(G', G)

---

tree. The algorithm is a parallel sweep with two types of events: (edge) contraction and splitting events. At a contraction event, the algorithm is invoked, recursively, on the resulting contour with fewer vertices. Besides tracking the next pair of bisectors that will cross (to detect contraction events), the algorithm also maintains pairwise distances between all the vertices of the polygon to detect a *splitting event*. Here, the segment between the pair of vertices now at the critical distance is used to divide the current contour polygon into two, and the algorithm is then invoked recursively on both. The merging occurring in the end extends the recursively computed crease pattern with an "annulus tiling" between two Lang polygons.

Splitting Graph. Additionally, in parallel we maintain a drawing of an outerplanar graph called a *splitting graph* as a book-keeping device for recording splitting events encountered during the algorithm. The initial splitting graph is equivalent to the input polygon. Each time a splitting event occurs, an edge is added to the splitting graph: when a splitting event occurs between corner vertices u and v the edge (u, v) is added to the splitting graph. See Fig. 4 for two example UMs and the associated splitting graphs.

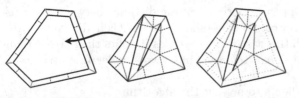

Fig. 8. A UM-tiling obtained by merging the crease pattern of the first contour into the tiled annulus

Main Data Structure: The UM-Tiling. The algorithm maintains a partial universal molecule (UM-tiling) of the contour representing the state of the parallel sweep at the current height h. When it recurses, this partial UM-tiling state is saved on a stack and the current contour is passed as a parameter to the new call.

When it returns from a recursive step, the algorithm *refines* the tiling on top of the stack with the information returned from the recursive call, which is a complete tiling of a sub-molecule, by subdividing the convex face of the contour on which the call was made with the returned UM tiling. See Fig. 8.

Types of Vertices and Edges in the UM-Tiling. The vertices, edges and faces of the tiling carry several attributes.

The edges of the UM-tiling are of several types: **polygon edges** and **splitting edges (solid black)**, **contour edges (dotted gray)**, parallel to a *parent* polygon or to a splitting edge, and **trace edges: solid gray** for the **bisectors** tracing the corner vertices of the polygon (those that correspond to the leaves of the tree), and **dashed black** for the **perpendiculars** tracing the marker vertices on the edges of the polygon (those corresponding to the internal nodes of the tree). The edge decorations in Fig. 9 make it easy to notice several connectivity patterns: the dashed black edges connect vertices labeled with internal tree nodes (b_i); the solid gray edges follow paths from leaf-vertices (a_i) that meet at vertices which, generically, are incident to 3 dashed black and 3 solid gray edges, in alternating circular order; the dotted gray edges of a convex contour are totally nested inside lower level contours. The (solid black) splitting edges partition a contour into smaller convex contours.

The UM-data structure also retains height information for all nodes (both as an index, indicating the nesting of the recursive call which computed them, and as a value); and pointers to *parent* nodes along directed paths towards the original polygon vertices; and several labels of these nodes. This information is used to prove invariant properties, including those mentioned briefly in this overview. The proof is by induction, with the base cases handling small polygons corresponding to trees with $1, 2$ and 3 leaves.

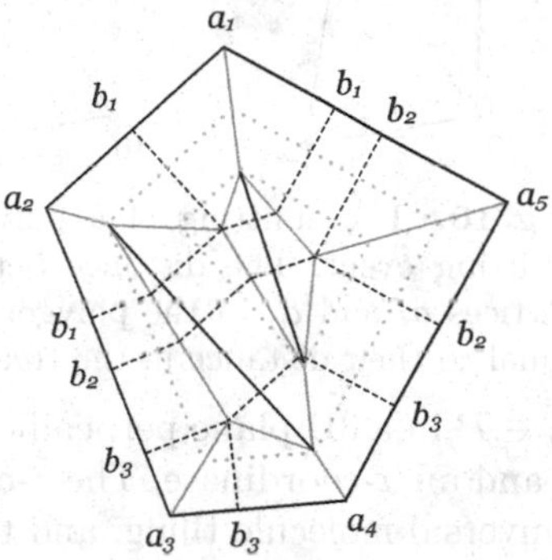

Fig. 9. A UM-tiling. The edge decorations denote the four types of edges: perpendiculars (dashed black), bisectors (solid gray), splitting (solid black) and contour edges (dotted gray).

Tiling the annulus is done by taking all bisectors of all the nodes (including the markers) of the current polygon (as in Fig. 8(a)), and extending them to the next contour based on a given height value, thus creating the vertices and edges of the next contour. This step also fills in the attributes of edges and vertices.

Polygon Splitting. To maintain recursively the invariant that $P_T(h)$ is a Lang polygon for $T(h)$, at a splitting event (corresponding to a pair of polygon vertices a_i and a_j), $P_T(h)$ is split into P_1 and P_2 by a diagonal from $a_i(h)$ to $a_j(h)$. The tree $T(h)$ is split into two trees T_1 and T_2, with T_1 defined as the nodes corresponding to the vertices traversed by a ccw walk from $a_i(h)$ to $a_j(h)$ in $P_T(h)$, and T_2 by a walk from $a_j(h)$ to $a_i(h)$ in $P_T(h)$. This splits the tree T along the path from a_i to a_j. Markers to the internal nodes of T_1 and T_2 along this path are maintained. Fig. 10 illustrates the splitting on a Lang polygon and its corresponding tree.

Merging Split Molecules. Finally, the call on line 14 of the algorithm merges the tilings from recursive calls after a splitting event, to form a tiling for the joint contour. Because there are matching copies of internal vertices along the sides in each polygon, the merge step just glues these copies back together. In terms of Lang bases, this step glues two recursively computed Lang bases along their common boundary to form the Lang base of the larger, merged contour.

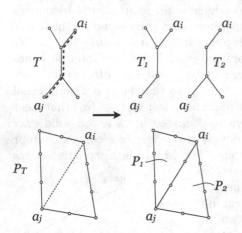

Fig. 10. The splitting of a polygon at a splitting event. The distance between the vertices a_i and a_j in the polygon becomes equal to their distance in the tree.

Merging the Border. To complete the UM-tiling for a Lang polygon, the UM-tiling of an inner contour is glued to the annulus tiling (previously computed) (Fig. 8). Because a UM-tiling does not split the contour edges, the edges of the polygon $P_T(h)$ are common to both the annulus and the crease pattern obtained recursively for $P_T(h)$.

Lang Base. We relate now properties of the universal molecule tiling for a Lang polygon to the second output of the algorithm, the flat Lang base. This is also recursively constructed (at least conceptually) by the algorithm. We define a mapping $B : V \mapsto P \subset \mathbf{R}^3$ of the vertices $i \in V$ to points $p_i \in P$ in a 3D plane perpendicular to the xy-plane. We need two coordinates: a z and an x-coordinate. The z-coordinate is the height value of the vertex in the universal molecule tiling, and the x is the coordinate of the corresponding node in the standard alignment of the tree. See Fig. 7(b).

The following invariants relate the data structures used in the algorithm.

Invariant 1 (Tree and Contour): The contour $P(h)$ at the time h of an event is a Lang polygon of the corresponding shrunken tree $T(h)$ obtained by reading off the edge length information from the vertex markings on the contour.

Invariant 2 (UM-Tiling and Lang Base): The UM-tiling $G(h)$ computed by the algorithm for a Lang polygon $R(h) = (T(h), P_T(h))$ satisfies the following properties: (1) the polygon boundary is preserved (i.e. no new vertices are added to it); (2) there exists a Lang base $B(h)$, isomorphic and isometric to the UM-tiling, which projects to the standard alignment of the tree $T(h)$ and maps the edges of $P_T(h)$ exactly onto their corresponding edges in the standard alignment of $T(h)$; and (3) the image of the contour $P(h)$ in the base $B(h)$ is the intersection of $B(h)$ with the $z = h$ plane.

This completes the high-level description of Lang's Universal Molecule algorithm.

5   Rigid Universal Molecules

In this section we prove Theorem 2 by showing that the universal molecule pattern from Fig. 3 is rigid. The main challenge is to prove rigidity in the absence of infinitesimal rigidity. Indeed, infinitesimal rigidity would have implied rigidity, but this is not the case here: an infinitesimal motion, with vertex velocities perpendicular to the plane of the "paper", always exists. For the proof, we introduce a different technique, called *rigidity transport*. It can be applied on any graph as long as it has vertices with 4 "unvisited" edges that act as "transmitters" (cf. definition given below) and which are reachable from a starting point via "transport" edges.

Single-Vertex Origami with 4 Creases. The faces surrounding a vertex incident to 4 edges (creases), isolated from the rest of the origami pattern, is called a 4-*edged single-vertex origami*. A generic realization of a 4-edged single-vertex origami is flexible, with a one-dimensional configuration space; this implies that when one of the dihedral angles is changed continuously, all the other dihedral angles are determined. We say that the origami has a one-dimensional *flex*. The flat open configuration is however not generic (it is singular), and thus may allow flexes to proceed along different branches of the one-dimensional configuration space. We rely on the tabulation of all the types of configuration spaces for planar 4-gons, which can be found for instance in [11], and on the relationship between the Euclidean, spherical and single-vertex origamis, as discussed at large in [17,16]. We start by identifying in Fig. 11 the types of single-vertex origamis with 4-creases (called, for simplicity, "4-edged gadgets") that appear in a Universal Molecule crease pattern. Notice their grouping into three categories, identified by the color of the central vertex: gray, white and black.

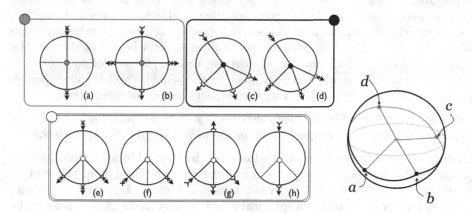

Fig. 11. *Left:* Rigidity transport of flat 4-edge single-vertex origamis appearing in a universal molecule. Arrows represent "input" and "output" edges. A cross on an arrow's tail indicates the edge remains open-flat, while a small crescent indicates a flex. An edge for which the behavior is not forced by the input edge is dotted. *Right:* The spherical four-bar mechanism *abcd* induced by a 4-edge single-vertex origami.

Rigidity Transport. Assume that a dihedral (input) edge of one such 4-edge gadget is kept, rigidly, in the flat open position (of 180°) or flexed by a small angle: can the behavior of the other (output) edges be predicted? For 4-edge gadgets created by the universal molecule algorithm and needed in the main proof of this section, Fig. 11 tabulates all the possibilities. The patterns are grouped into three categories: black, white, and gray. In each category opposite angles are supplementary (i.e. $\alpha + \gamma = \beta + \delta = 180°$). In Fig. 11 the category of each gadget is indicated by the shading of the center vertex. An arrow pointing towards the center indicates the "input" to the gadget. For easy reading we denote the "signals" by a marker on the tail of the arrow: a cross indicates that the dihedral edge is kept at 180°, and a small crescent indicates a slight perturbation (flex). An arrow pointing outwards indicates a forced behavior on another edge. A dotted edge signifies that its behavior is not determined by the input, and that it can either stay flat or have a small flex.

In **white**, one pair of opposite edges are aligned and the other two make equal angles (different from 90°) with them. In **gray**, both pairs of opposite edges are aligned, and one is perpendicular to the other (i.e. all face angles are 90°). In **black**, opposite angles are supplementary. We analyze these patterns with respect to an "input" crease, i.e. the mechanical action of keeping the dihedral angle as it is or perturbing it slightly. In each case if an edge is flexed resp. remains rigid, then its opposite edge is flexed resp. remains rigid; in addition, in **black**, all edges either flex or remain rigid collectively (Fig. 11, c, d); in **white** if an aligned edge remains rigid (Fig. 11,e) or an unaligned edge flexes (Fig. 11,g), then all edges remain rigid or flex (resp.); and in **gray** if an edge flexes, then the opposite pair of edges remain rigid (Fig. 11,b).

These categories will be used as follows: (a) the black gadgets will appear at the endpoints of a splitting edge, (b) the white gadgets to a contraction event and (c) the gray edges will apply along a splitting edge, at the marker vertices present along it. Here keeping one edge rigid, resp. deforming it slightly, forces its aligned pair to have the same behavior; moreover, the deformation of the dihedral angle of an edge forces the flatness (and hence, rigidity) of the perpendicular pair. This process (of inferring rigidity/flatness of an edge from what happens with another edge incident to the same vertex) is called *rigidity transport*. We summarize these very simple facts in the following Lemma.

Lemma 1. *The rigidity/flexibility dependency patterns from Fig. 11 correctly depict 4-vertex origami configurations in the vicinity of the flat-open state.*

Proof. The proof relies on a calculation made by Bricard in his famous memoir [5] on flexible octohedra; see [6] for a modern English translation. What in [6] is termed a *tetrahedral angle* is, in our terms, a 4-edged single-vertex origami, and can further be viewed as a spherical four-bar mechanism. A *face angle* is the interior angle of the central vertex, and a *dihedral angle* is the angle between the planes supporting two adjacent faces. Bricard's memoir has two parts. The first analyzes the relationship between adjacent dihedral angles of a spherical four-bar mechanism. The second is his analysis of flexible octohedra, which we do not use. Bricard categorizes the spherical four-bar mechanisms into three types based on relationships between the face angles. All of our gadgets are of Bricard *Type 3b*, which comprises those 4-edged single-vertex origami in which opposite angles are supplementary.

Given a gadget from Fig. 11, let face angles $\alpha, \beta, \gamma = \pi - \alpha, \delta = \pi - \gamma$ be given in (cw or ccw) order and let ϕ and ψ be the dihedral angles between the two faces spanning the (α, β), resp. (β, γ) pairs of angles. The angle labels are illustrated in Fig. 11(b, c, e, f). The dihedral angle values are taken between 0 and π. Let $t = \tan(\phi/2)$ and $u = \tan(\psi/2)$. Bricard showed that for *Type 3b* structures:

(1) opposite dihedrals have equal angle measure, and

(2) the parameters t and u satisfy the quadratic equation:

$$sin(\alpha + \beta)t^2 + 2\sin\beta tu - \sin(\alpha - \beta)u^2 = 0$$

Equation (2) is valid unless ϕ or ψ are π. In that case we need a different derivation. If instead we define t and u by $t = \cot(\phi/2)$ and $u = \tan(\psi/2)$ and following Bricard's derivation we obtain:

(2') the parameters t' and u' satisfy the quadratic equation:

$$sin(\alpha - \beta)t^2 + 2\sin\beta tu - \sin(\alpha + \beta)u^2 = 0$$

In Bricard's derivation, $\tan\beta$ and $\tan\delta$ terms appear, which means that it will not work for cases (a) and (b), since the tangent is undefined for $\beta = \delta = \pi/2$. To prove cases (a) and (b) we use elementary spherical geometry. To prove (c)-(h) we use (1) and (2') above. The rigidity implications of (f) and (h) follow immediately from property (1). We next prove cases (c-e) and (g) using (2'). We then derive cases (a) and (b) with elementary spherical geometry.

In the **black** and **white** cases (c-e) and (g), we have the sum of the opposite angles $\alpha + \gamma = \beta + \delta = \pi$. Assume without loss of generality that $0 < \beta \le \alpha < \pi$ and $\alpha + \beta \ne \pi$. In this case the coefficients in equations (2') are not zero. A simple analysis of these two equations shows that $t = 0$ if and only if $u = 0$. This completes the proof of all four cases.

We prove the **gray** cases (a) and (b) using the corresponding spherical four-bar linkage $abcd$ on the unit sphere where a corresponds to the "input" crease (See Fig. 11(right)). The length of each bar is $\pi/2$ and the interior angle of each corner is equal to the dihedral of the corresponding crease. In case (a), $\angle a$ is π and so b and d are antipodal. By elementary spherical trigonometry, $\angle c$ must be π as well. In (b) we assume that $\angle a$ is between 0 and π. We add a bar bd to form spherical triangles abd and cbd. By elementary spherical trigonometry it follows that $\angle b$ and $\angle d$ in abd and cbd must be $\pi/2$ completing the proof.   □

We are now ready to prove:

Theorem 2. (Rigid Universal Molecules) *There exist universal molecules which are rigid in the open-flat state.*

Proof. Overview: We prove existence by showing that the universal molecule crease pattern in Fig. 3 is rigid. The proof is by contradiction: we assume that the crease pattern is flexible and derive a contradiction by analyzing a potential nearby realization, in which the dihedral angle of at least one edge must be (slightly) smaller than 180°. We use the types of folds that may occur at vertices of degree 4, as classified in Lemma 1. We will start at the central vertex and "propagate" flat (rigid) edges and flexed edges by sequentially applying one of the inferences (input-implies-output) proven in Lemma 1 and illustrated in Fig. 11. If an edge incident to a vertex is rigid, the degree of the vertex is reduced by one, when its flexibility is analyzed. A step in such a sequence of inferences is illustrated in Fig. 12.

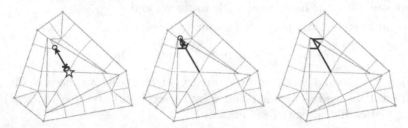

Fig. 12. An example of the logical inference used to prove rigidity (Case 1). The star vertex at the center is assumed to be rigid. We analyze the circled vertex in each figure and conclude, from its type and rigidity of one incident edge, the rigidity of its neighbors. We continue this process for all vertices and conclude that the entire origami must be flat.

The analysis of flexibility of our example reduces to just two cases. *First,* we assume that at the most central vertex of the crease pattern, all edges remain flat. Using Lemma 1, we then iteratively propagate the "flatness" and infer that all edges must remain flat. *Otherwise,* one of the edges incident to the central vertex is not flat, i.e. either a valley or a mountain. Following again the simple rules of local foldability at neighboring vertices, we arrive at a contradiction where one edge will need to be both flat, and not flat, simultaneously. From this we conclude that the crease pattern in Fig. 3 should be rigid.

We analyze now in detail each case, using the guidelines from Fig. 13: (left) vertex labels, (center) vertex types (black, gray and white, as classified in Lemma 1 and illustrated in Fig. 11) and (right) an oriented inference "path" leading to a contradiction, in Case 2 below.

Case 1: All Edges Incident to 1 Are Flat. The following inference (see Fig. 12) show that, in this case, all edges of the crease pattern have to be flat: the vertices 2, 6, 10, incident to 1, are white, have a rigid input edge (the one coming from 1), hence the other edges (to 5, 7, 18 etc.) are implied to be rigid; 13, 18, 22 are black each with a flat incident edge, and thus all edges incident to them are flat. Therefore, 17, 23, 28 (also black) are flat. Then the gray 3, 5, 7, 9, 12, and 24 are incident to two flat edges incident to the same face and are thus flat. By the same reasoning, 4, 8 and 11 are flat; 16, 21, and 27 are white, with one flat edge, hence all are flat; 15, 20, 26 are flat by the same reasoning; 14, 19, 25 now have all but 3 edges proved to be flat by the statements above;

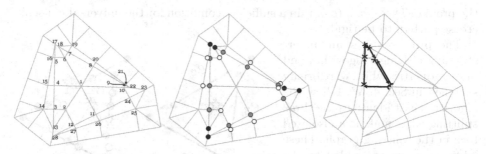

Fig. 13. Illustration of the methodology for deriving a contradiction from the assumption that this crease pattern flexes

since none of these are collinear, they must be all flat. Thus all edges are flat: contradiction.

Case 2: Some Edge Incident to Vertex 1 Is Not Flat. In this case, at least 4 of them must be non flat, and hence in one of the 6 pairs of consecutive edges, neither edge is flat. A contradiction will be derived in each case, and all cases are similar, so we present the argument only for the pair of edges (1,4) and (1,6), assumed to be displaced (not flat), as in Fig. 13(right). Since (1, 4) is not flat and vertex 4 is gray, edge (4, 5) is flat; since (4, 5) is flat, and 5 is gray, then (5, 17) is flat; (5, 17) flat and 17 black, implied that (17, 18) is flat; (17, 18) being flat implies that (18, 6) is flat. Finally, (18, 6) being flat implies by (b) that (6, 1) is flat. This contradicts the assumption that (1, 6) is deformed away from flatness. What completes the proof is the observation that the same sequence of inferences applies to all possible subsets of edges incident to 1.   □

6   Crease Patterns with Isolated Peaks

The distinguishing feature of the degree-6 vertex in the example used to prove Thm. 2 is that each edge incident to it is also incident to a splitting edge and not a boundary edge. We call such a vertex an *isolated peak* of the crease pattern. A degree-6 isolated peak occurs if and only if the splitting graph (defined in Sec. 4) has an internal triangular face. We now extend the result of the previous section to all generic universal molecule crease patterns with an isolated peak.

Theorem 3. (Universal Molecules with isolated peaks are rigid)*If a generic crease pattern produced by Lang's Universal Molecule algorithm has an isolated peak, then it is rigid in the open-flat state.*

Proof. The intuition behind the proof comes from the following observation: the example shown to be rigid by Theorem 2 contains a special vertex "surrounded" by three split edges. On the other hand, the leftmost molecule from Fig. 4 doesn't have such a vertex. We use now the splitting graph, defined previously as an outerplanar graph whose outer cycle corresponds to the given polygon and the diagonals correspond to the splitting events. Then we apply the argument used in the proof of Theorem 2 to obtain a sufficient condition for the universal molecule crease pattern to be rigid.

The proof requires an under-standing of Lang's algorithm and of its properties which were briefly sketched in Sec. 4. We follow the algorithm as it identifies the three splitting edges making an isolated face in the splitting graph. These edges e_1, e_2, e_3 are added to the Universal Molecule crease pattern at events happening at different heights $h_1 < h_2 < h_3$ (by the as-sumption of genericity). Then, the

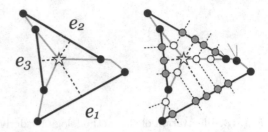

Fig. 14. The crease pattern induced by an iso-lated peak in the splitting graph

splitting edge e_1 is on the h_1-contour, and includes edge e_2 and its h_2-contour, which in turn includes edge e_3 and its h_3-contour.

Extending the crease pattern with bisector and perpendiculars around these three splitting edges, we obtain a pattern illustrated in Fig. 14. On the left, we see the three splitting edges (solid black) and their endpoints (of "black"-type, according to the classification from Lemma 1). There is a unique center vertex (the "peak") of degree 6 denoted by a star, with three solid gray bisectors and three dashed black perpendiculars emanating from it. The other endpoints of the solid gray bisectors are exactly as they are depicted in the picture: two go to edge e_3, the latest to be added as a split edge, and one goes to e_2. The endpoints of the splitting edges are connected by a path of bisector edges, and additional vertices of the crease pattern may be present along all these segments, as illustrated in Fig. 14(right).

With this pattern in place, one recognizes immediately the applicability of the proof of Theorem 2 to derive that all the edges that are part of this figure (of three splitting segments and all of their incident edges) must be rigid. To complete the proof for the entire crease pattern, we proceed by induction. First, we identify a few properties of the Universal Molecule crease pattern (which follow from the invariants of the algorithm). The base case of a generic recursive call to Lang's algorithm is when the contour polygon is a triangle. The bisectors of the triangle meet in one vertex, which we'll call a peak; indeed, in the Lang base state, these will be points of local maximum height for the folded paper. Generically, these are the only vertices of degree larger than 4 (namely, 6) that appear in the crease pattern. Next, we remark that each splitting segment has exactly one peak vertex on each side, and each peak is connected by two paths of bisector segments, to the endpoints of each split segment in its vicinity and by a path of perpendicular edges to some point on such a splitting edge. Therefore, if a splitting edge is proven to be "rigid" because of what happens on one of its sides, then all the edges incident to it are so, and the rigidity is transported to the peak on the other side. To complete the proof, we show inductively (proceeding outwards from an inner triangle towards the polygon sides) that all the split edges become rigid, once those of an isolated triangle (in the split graph) have been proven to be so.   □

7   Stable Lang Bases

We now extend the previous arguments to prove that not only is the flat state inflexible, but the folded Lang base state produced by Lang's algorithm is *stable* and cannot be unfolded. This requires first some conceptual clarifications.

The Lang Base State. In this state, the *perpendicular* creases of the universal molecule are grouped together, overlapping in groups that project to internal nodes of the input metric tree. They act as hinges about which *flaps* can be rotated. The remaining creases are folded completely as either mountain or valley folds.

The Lang base state is therefore flexible, suggesting that it may be able to reach other interesting configurations through appropriate deformations. Two faces sharing creases that are bisectors or splitting edges are folded flat, one on top of the other, while the flaps made of faces sharing a perpendicular edge will have a rotation motion. Fig. 15, showing from below a slight perturbation of a Lang base (just enough to see which faces overlap and which not) may help with visualizing these properties. A state which is obtained from the flat Lang base simply by rotating the flaps about their incident hinges is said to be *tree-reachable*.

We prove now that the Lang base state may sometimes be stable, or *not unfoldable*, meaning that there is no nearby configuration which is not tree-reachable. To be *unfoldable*, i.e. to unfold, requires the bisector and splitting edges to *open* slightly. Our goal is to show that no such crease is opening, i.e. it cannot have a non-zero dihedral angle (while maintaining rigidly the faces) in a small neighborhood of some tree-reachable configuration.

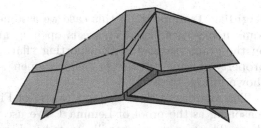

Fig. 15. A Lang base (view from below), visualized by slightly perturbing the metric while maintaining the combinatorial structure of the realization, so that its folding pattern can be seen

Theorem 4. (Stability of Lang Bases with isolated peaks) *If a generic crease pattern produced by Lang's Universal Molecule algorithm has an isolated peak then the Lang base associated to it cannot be unfolded, in the sense that overlapping panels grouped into flaps cannot be separated.*

Proof. We follow a similar plan as for Theorem 3. The critical step is the base case, i.e. the counterpart of Theorem 2, which relies on a slightly different set of gadgets. These, and the chain of implications leading to a contradiction to the assumption that the base is not stable in one of two cases, are depicted in Fig. 16. The generalization to those cases where the splitting graph has internal triangles is the same as in Theorem 3, therefore we focus now on the base case.

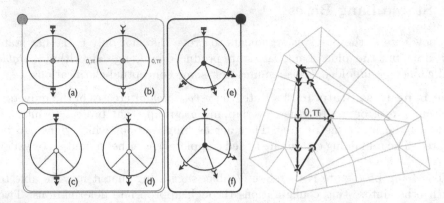

Fig. 16. The gadgets used in the proof of Theorem 4, and the chain of implications in one of two cases needed to contradict the hypothesis that an unfolding of the Lang base exists. Arrows indicate input and output edge behavior. One with two bars indicates a "folded" (i.e. dihedral angle zero) edge. One with a crescent indicates an "opened" (i.e. dihedral angle slightly larger than zero) edge. Dotted edges are undetermined and gray edges are a special type in which the dihedral may be only either 0 or 180°.

Rigidity Transport. In this case we assume that a nearby state exists where some non-perpendicular crease is opening and analyze the rigidity transport on the graph. Instead of transporting "flatness" of an edge, we transport the property of being "closed" or "slightly open". The gadgets for this transport are shown in Fig. 16 (left).

The analysis of these gadgets (a)-(f) of Fig. 16 essentially follows the same reasoning as the proof of Lemma 1. We use the same results of Bricard listed there. The transport cases illustrated in Figure 16 (c), and (d) follow directly from (1). The proof of (a) follows from the observation that if $\angle a = 0$ in the spherical four-bar linkage $abcd$ induced by the origami, then $b = d$ and the spherical triangle bcd has a zero-length edge bd. The proof of the transport of (b) is the same as the proof of the transport of (b) in Fig. 11. The proof of the transport of (e) and (f) is exactly the same as for (c) and (d) of Fig. 11 except using equation (2) rather than (2').

We now complete the proof of Theorem 4. To obtain a contradiction, we assume that a nearby state exists and proceed in two cases: **1.** all non-perpendiculars incident to 1 are closed, and **2.** one of the non-perpendiculars is opening. In each case we derive a contradiction. We analyze in detail each case, using the vertex labels (left) and vertex types (center) from Fig. 13, and the oriented inference "path" from Fig. 16 (right) leading to a contradiction in Case 2 below.

Case 1: All Non-perpendiculars Incident to 1 Are Closed. The proof of this is exactly the same as **Case 1** for Theorem 2 except that, rather than flatness, the property of an edge of being closed is propagated. In order to obtain the contradiction we do not need to propagate along perpendiculars. This is because the propagation rules for closed and flat edges are the same for the vertex types we encounter here along the non-perpendiculars.

Case 2: Some Non-perpendicular Incident to 1 Is Opening. We will show that the three non-perpendiculars incident to one are opening, and the incident perpendiculars are each *either closed or flat*. Since this state of a degree 6 vertex does not exist in the general case in which the angles at 1 are not all equal, this is a contradiction. We will assume that $(1, 6)$ is opening, and show the inference necessary to prove $(1, 2)$ is opening and $(1, 4)$ is 0 or $180°$. Since $(1, 6)$ is opening and 6 is white, $(6, 18)$ is opening. Since $(6, 18)$ is opening, and 18 is black, $(18, 17)$ is opening. Similarly, $(17, 5)$ is opening. $(17, 5)$ is opening, and 5 is gray, so $(5, 4)$ is opening. $(5, 4)$ is opening, and 4 is gray so $(4, 3)$ is opening, *and* $(4, 1)$ is either 0 or $180°$. $(4, 3)$ is opening, and 3 is gray, so $(3, 13)$ is opening. $(3, 13)$ is opening, and 13 is black, so $(13, 2)$ is opening. $(13, 2)$ is opening, and 2 is white so $(2, 1)$ is opening. To complete the proof, we observe that an analogous sequence beginning with $(1, 2)$ shows that $(1, 11)$ is 0 or $180°$ and $(1, 10)$ is opening. Then the same sequence beginning at $(1, 10)$ shows $(1, 8)$ is 0 or $180°$. This does not depend on which non-perpendicular of 1 we begin with and no such state exists for vertex 1.   □

8   Conclusion

For the family of crease patterns generated by Lang's Universal Molecule algorithm, in which the outerplanar splitting graph has an isolated peak, we have proven that the final Lang base state is not reachable by continuous flat-faced folding from the initial flat state. Even stronger, we showed that the initial, creased paper does not move at all if the faces are to remain rigid. We also proved that for the same crease patterns, the folded Lang base state cannot be unfolded.

Fig. 4 shows an example of a metric tree and two possible Lang molecules for it, whose splitting graphs indicate that one is rigid. The flat-face flexibility of the other, if true, will have to be established by other means. A full characterization of the flexible origami patterns produced by Lang's algorithm remains an open question, and Lang's algorithm requires further investigation as to which crease patterns yield continuously foldable origamis.

References

1. Bern, M.W., Hayes, B.: The complexity of flat origami. In: SODA: ACM-SIAM Symposium on Discrete Algorithms (1996)
2. Bern, M.W., Hayes, B.: Origami embedding of piecewise-linear two-manifolds. Algorithmica 59(1), 3–15 (2011)
3. Bowers, J.C., Streinu, I.: Lang's universal molecule algorithm (video). In: Proc. 28th Symp. Computational Geometry, SoCG 2012 (2012)

4. Bowers, J.C., Streinu, I.: Lang's universal molecule algorithm. Technical report, University of Massachusetts and Smith College (December 2011)
5. Bricard, R.: Mémoire sur la théorie de l'octaèdre articulé. J. Math. Pure et Appl. 5, 113–148 (1897)
6. Bricard, R.: Memoir on the theory of the articulated octahedron. Translation from the French original of [5] (March 2012)
7. Demaine, E.D., Demaine, M.L.: Computing extreme origami bases. Technical Report CS-97-22, Department of Computer Science, University of Waterloo (May 1997)
8. Demaine, E.D., Demaine, M.L., Lubiw, A.: Folding and one straight cut suffices. In: Proc. 10th Annual ACM-SIAM Sympos. Discrete Alg. (SODA 1999), pp. 891–892 (January 1999)
9. Demaine, E.D., O'Rourke, J.: Geometric Folding Algorithms: Linkages, Origami, and Polyhedra. Cambridge University Press (2007)
10. Eppstein, D.: Faster circle packing with application to nonobtuse triangulations. International Journal of Computational Geometry and Applications 7(5), 485–491 (1997)
11. Farber, M.: Invitation to Topological Robotics. Zürich Lectures in Advanced Mathematics. European Mathematical Society (2008)
12. Lang, R.J.: A computational algorithm for origami design. In: Proceedings of the 12th Annual ACM Symposium on Computational Geometry, pp. 98–105 (1996)
13. Lang, R.J.: Treemaker 4.0: A program for origami design (1998)
14. Lang, R.J.: Origami design secrets: mathematical methods for an ancient art. A.K. Peters Series. A.K. Peters (2003)
15. Lang, R.J. (ed.): Origami 4: Fourth International Meeting of Origami Science, Mathematics, and Education. A.K. Peters (2009)
16. Panina, G., Streinu, I.: Flattening single-vertex origami: the non-expansive case. Computational Geometry: Theory and Applications 46(8), 678–687 (2010)
17. Streinu, I., Whiteley, W.: Single-vertex origami and spherical expansive motions. In: Akiyama, J., Kano, M., Tan, X. (eds.) JCDCG 2004. LNCS, vol. 3742, pp. 161–173. Springer, Heidelberg (2005)
18. Wang-Iverson, P., Lang, R.J., Yim, M.: Origami 5: Fifth International Meeting of Origami Science, Mathematics, and Education. Taylor and Francis (2011)

Algebraic Analysis of Huzita's Origami Operations and Their Extensions

Fadoua Ghourabi[1], Asem Kasem[2], and Cezary Kaliszyk[3]

[1] Kwansei Gakuin University, Japan
ghourabi@kwansei.ac.jp
[2] Taylor's University, Malaysia
asem.kasem@taylors.edu.my
[3] University of Innsbruck, Austria
cezary.kaliszyk@uibk.ac.at

Abstract. We investigate the basic fold operations, often referred to as Huzita's axioms, which represent the standard seven operations used commonly in computational origami. We reformulate the operations by giving them precise conditions that eliminate the degenerate and incident cases. We prove that the reformulated ones yield a finite number of fold lines. Furthermore, we show how the incident cases reduce certain operations to simpler ones. We present an alternative single operation based on one of the operations without side conditions. We show how each of the reformulated operations can be realized by the alternative one. It is known that cubic equations can be solved using origami folding. We study the extension of origami by introducing fold operations that involve conic sections. We show that the new extended set of fold operations generates polynomial equations of degree up to six.

Keywords: fold operations, computational origami, conic section.

1 Introduction

Origami is commonly conceived to be an art of paper folding by hand. It is not restricted to an art, however. We see continuous surge of interests in scientific and technological aspects of origami. It can be a basis for the study of geometrical systems based on reflections, and the industrial applications can be found abundantly, such as in automobile industry, space industry etc.

In this paper, we focus on the algebraic and geometrical aspects of origami. Suppose that we want to construct a geometrically interesting shape, say a regular heptagon, from a sheet of paper. We need a certain level of precision even though we make the shape by hand. Since we do not use a ruler to measure the distance, nor do we use additional tools[1], what we will do by hand is to construct creases and points. The creases are constructed by folding the paper. The creases we make are the segments of the lines that we will treat. The creases and the four edges of the initial origami form line segments. The points are constructed

---

[1] This restriction will be relaxed in Section 7.

T. Ida and J. Fleuriot (Eds.): ADG 2012, LNAI 7993, pp. 143–160, 2013.
© Springer-Verlag Berlin Heidelberg 2013

by the intersection of those segments. In the treatment of origami[2] in this paper, origami is always sufficiently large such that, whenever we consider the intersections of segments, we consider the intersections of the lines that extend the segments. The shape of the origami that we want to obtain by folding is a (possibly overlaid) face(s) constructed by the convex set of the thus-constructed points.

As the crease is constructed by folding, the main question is how to specify the fold. Since we fold an origami along the line, the question is boiled down to how to specify the line along which we fold the origami. We call this line *fold line*.

In 1989 Huzita proposed the set of basic fold operations often referred to as Huzita's axiom set [4]. Later studies showed that Huzita's set of fold operations is more powerful than Euclidean tools, i.e. straightedge and compass (abbreviated to SEC hereafter), in that we can construct a larger set of points of coincidence by applying Huzita's set of operations than by SEC [1]. More precisely, the field of origami constructible numbers includes the field of SEC constructible numbers, therefore, the class of the shapes formed by connecting the coincidences is richer than that of the shapes formed by SEC. The trisector of a given arbitrary angle is a famous example that is constructible by origami, but not by SEC [11,6]. This triggered the activities of researchers who are interested in mathematical aspects of origami such as the contribution of Martin [10] and Alperin [1]. Although several studies have been made to confirm the power of origami as we have seen above, we propose a more rigorous treatment of Huzita's set of operations. We choose not to use the terminology of Huzita's *axiom* itself as the set does not constitute a mathematical axiom set. The need for a formal method that can serve as a rigorous standard for the origami theory is pressing since folding techniques have been adapted in industry. This paper presents a preparatory but necessary rigorous statements of Huzita's basic fold operations towards the formalization of origami theory in proof assistants.

In this paper we restate Huzita's basic fold operations. We make the new statements more precise by clarifying the conditions that enable folds. We analyze the operations algebraically and present theorems about the finite number of fold lines. We also introduce a general origami principle that performs all the operations. Furthermore, we extend the capability of basic fold operations by introducing conic sections and show that this extension is defined by equations of degree six.

The structure of the rest of the paper is as follows. In Section 2, we summarize our notions and notations. In Section 3, we present Huzita's basic fold operations. In Section 4, we define the possible superpositions of geometrical objects of origami. In Section 5, we reformulate the basic fold operations. In Section 6, we introduce a general origami principle that performs all the basic fold operations. In Section 7, we consider superpositions of points and conic sections. In Section 8, we conclude with remarks on future directions of research.

---

[2] The word *origami* is also used to refer to a square sheet of paper used to perform origami.

2 Preliminaries

An origami is denoted by $\mathcal{O}$. An origami $\mathcal{O}$ is supposed to represent a square sheet of paper with four points on the corners and four edges that is subject to folding. Some intersections of lines may not fit on the square paper. However, we want to work with these points. To achieve this, we consider $\mathcal{O}$ to be a sufficiently large surface so that all the points and lines that we treat are on $\mathcal{O}$.

In this paper, we restrict the use of geometrical objects only to points, lines and s-pairs (to be defined in Section 4). We use α and β to note either a point or a line. Points are denoted by a single capital letter of the Latin alphabet such as A, B, C, D, P, Q etc.[3], and lines are denoted by γ, k, m, and n. Since we use Cartesian coordinate system in this paper, a line is represented by a linear equation $ax + by + c = 0$ in variables x and y. The notation "$f(x, y) :=$ *polynomial* in x and $y = 0$" is used to declare that f is a curve represented by the equation *polynomial* $= 0$. (x, y) on the lefthand side of $:=$ may be omitted.

The sets of all points and lines are denoted by Π and $\mathcal{L}$, respectively. Abusing the set notation we use $P \in m$ to mean point P is on line m.

For a set $\mathcal{S}$, we notate its cardinality by $|\mathcal{S}|$. For two lines m and n, $m \parallel n$ is true when m and n are parallel or equal.

3 Fold Principle

3.1 Basic Idea

By hand we can fold the origami by tentatively making a line either to let it pass through two points or to superpose points and lines. The former corresponds to applying a straightedge in Euclidean construction. In practice, to construct a line that passes through a point we bend the paper near the point and we flatten the paper gradually until we make the point lie on the intended fold line. The latter is pertinent to origami. Superposition involves two points, a point and a line, and two lines. To superpose two points, we bring one point to another, and then we flatten the paper. To superpose a point and a line, the easy way is to bring the point onto the line, and then we flatten the paper. Superposition of two lines is more complex, and we will treat this operation along with its algebraic meaning in Section 4.

3.2 Huzita's Basic Fold Operations

We restate the set of seven basic fold operations of Huzita. The first six were proposed by Huzita and below are their statements as they appear in [4]. The seventh was proposed by Justin [7] and rephrased by us to fit in with Huzita's statements.

---

[3] $A \sim F$, X and Y are overloaded, in fact. The meaning the symbols denote should be clear from the context.

(1) Given two distinct points, you can fold making the crease pass through both points (ruler operation).
(2) Given two distinct points, you can fold superposing one point onto the other point (perpendicular bisector).
(3) Given two distinct (straight) lines, you can fold superposing one line onto another (bisector of the angle).
(4) Given one line and one point, you can fold making the crease perpendicular to the line and passing through the point (perpendicular footing).
(5) Given one line and two distinct points not on this line, you can fold superposing one point onto the line and making the crease pass through the other point (tangent from a point to a parabola).
(6) Given two distinct points and two distinct lines, you can fold superposing the first point onto the first line and the second point onto the second line at the same time.
(7) Given a point and two lines, you can fold superposing the point onto the first line and making the crease perpendicular to the second line.

We will call this set of basic fold operations *Huzita's fold principle*.

The set of points and lines that can be constructed using the first five and seventh operations is the same as the set of points and lines that can be constructed by SEC. The sixth operation is more powerful: it allows constructing common tangents to two parabolas which are not realizable by SEC.

Huzita and Justin carefully worked out the statements to exclude the cases that give infinite fold lines by imposing conditions on points and lines (e.g. distinct points, distinct lines, etc.). However, some of these conditions are insufficient or unnecessary. A thorough discussion on Huzita's statements is given in [9].

While these statements are suitable for practicing origami by hand, a machine needs stricter guidance. An algorithmic approach to folding requires formal definition of fold operations. Furthermore, we need to explicitly identify the conditions that ensure the finite number of fold lines.

4   Superposition

We define a superposition pair, *s-pair* for short, (α, β). It is a pair of geometrical objects α and β that are to be superposed. An s-pair (α, β) defines a fold line along which the origami is folded to superpose objects α and β. Depending upon the types of the objects, we have the following superpositions.

Point-Point Superposition. When points P and Q are distinct, the s-pair (P, Q) defines a unique fold line that superposes P and Q. This unique line is the perpendicular bisector of the line segment whose start and end points are P and Q, respectively, and is denoted by $P \updownarrow Q$

When points P and Q are equal, the s-pair (P, Q), i.e. the s-pair (P, P), does not define a unique fold line to superpose P onto itself. Points P and P are superposed by any fold line that passes through P. Namely, the s-pair (P, P) defines the infinite set $\mathcal{I}(P)$ of fold lines that pass through P, i.e.

$$\mathcal{I}(P) = \{\gamma \mid P \in \gamma\}$$

Here we note that two sets $\mathcal{I}(P)$ and $\mathcal{I}(Q)$ can define a line, denoted by PQ, passing through points P and Q, i.e. $\gamma = \mathcal{I}(P) \cap \mathcal{I}(Q)$. The straightedge operation can even be replaced by the superpositions.

Line-Line Superposition. When lines m and n are equal, what we do is superposing a line onto itself. This is achieved in the following way. Choose an arbitrary point on the line, divide the line into the two half lines, and then superpose the two half lines with the dividing point at the ends of the both half lines. Geometrically speaking, we construct a perpendicular to m and fold along that perpendicular. Any perpendicular to m superposes the line m onto itself. Hence, the s-pair (m, m) defines the following infinite set $\mathcal{B}(m)$ of fold lines.

$$\mathcal{B}(m) = \{X \updownarrow Y \mid X, Y \in m, X \neq Y\}$$

Note that, in passing, we exclude m itself from $\mathcal{B}(m)$. Namely, m is *not* considered as the fold line to superpose m onto itself as this does not create new lines.

To superpose two distinct lines, we assume the capability of hands that slides a point along a line. By the combination of superposition and of sliding, we can achieve the superposition of two distinct lines.

Point-Line Superposition. An s-pair (P, m) defines the following set $\Gamma(P, m)$ of fold lines that superpose P and m.

$$\Gamma(P, m) = \begin{cases} \{X \updownarrow P \mid X \in m\} & \text{if } P \notin m \\ \mathcal{B}(m) \cup \mathcal{I}(P) & \text{if } P \in m \end{cases}$$

Here we define $\Gamma(P, m)$ by cases of $P \notin m$ and $P \in m$. If $P \notin m$ then the fold line that superposes P and m is a tangent to the parabola with focus P and directrix m [7,10,1]. Therefore, $\{X \updownarrow P \mid X \in m\}$, in the former case, denotes the set of tangents of the parabola defined by the focus P and the directrix m. The latter corresponds to folding along any perpendicular to m or along any line that passes through P.

5   Formulation of Fold

5.1   Revisit of Huzita's Fold Principle

Table 1 shows the reformulation of Huzita's fold principle by a superposition or combinations of two superpositions. Each row of Table 1 corresponds to each basic operation given in Subsection 3.2. The second column shows the superpositions used to formalize each fold operation. The third column summarizes the degenerate cases of each operation. In practice, a degenerate case means infinite folding possibilities to achieve the superpositions in the second column. Huzita implicitly assumed $P \notin m$ whenever a point P and a line m are to be superposed. The fourth column indicates this assumption of the incidence relation. An incident case is one where the s-pair $(\alpha, \beta) \in \Pi \times \mathcal{L}$ has the property $\alpha \in \beta$. This can occur in the operations where we have point-line superposition(s), namely operations (5), (6) and (7). In the case of (6), it is enough to have only one s-pair that has the property $\alpha \in \beta$.

Propositions of incidence may cover some of degenerate configurations. In Table 2, redundancy of propositions is avoided in the last two columns. For instance in the case of (5), if $P = Q \wedge P \in m$ then there are infinite possible fold lines passing through Q and superposing P and m. More precisely, any line passing through P is a possible fold line. Or, the proposition $P \in m$ of incidence covers the degeneracy proposition. In other words, by eliminating the case where $P \in m$, we also eliminate the case where $P = Q$ and $P \in m$. The proposition $P = Q \wedge P \in m$ is removed from degeneracy column of the operation (5) in Table 2. The more general condition, i.e. $P \in m$, is kept in the incidence column.

Table 1. Superpositions in Huzita's fold principle

operation	s-pairs	degeneracy	incidence
(1)	$(P,P),(Q,Q)$	$P = Q$	—
(2)	(P,Q)	$P = Q$	—
(3)	(m,n)	$m = n$	—
(4)	$(m,m),(P,P)$	—	—
(5)	$(P,m),(Q,Q)$	$P = Q \wedge P \in m$	$P \in m$
(6)	$(P,m),(Q,n)$	$(P \in m \wedge Q \in n \wedge (m \parallel n \vee P = Q)) \vee$ $(P \notin m \wedge Q \notin n \wedge m = n \wedge P = Q)$	$P \in m \vee Q \in n$
(7)	$(P,m),(n,n)$	$m \parallel n \wedge P \in m$	$P \in m$

The notion of superposition enable us to reformulate Huzita's fold principle. We first introduce a function ζ that, given a sequence of s-pairs, computes all the fold lines that realize all the given s-pairs (i.e. superpose the elements). The detailed definition of ζ is beyond the scope of this paper. Function ζ has been implemented as the core of computational origami system Eos [3,5]. We provide the reformulation of Huzita's fold principle: a new set of operations that specify ζ. We denote this new formalization by **H**.

(O1) Given two distinct points P and Q, fold $\mathcal{O}$ along the unique line that passes through P and Q.

Table 2. Superpositions in Huzita's fold principle with simpler conditions for degeneracy

operation	s-pairs	degeneracy	incidence
(1)	$(P,P),(Q,Q)$	$P=Q$	–
(2)	(P,Q)	$P=Q$	–
(3)	(m,n)	$m=n$	–
(4)	$(m,m),(P,P)$	–	–
(5)	$(P,m),(Q,Q)$	–	$P \in m$
(6)	$(P,m),(Q,n)$	$P=Q \wedge m=n$	$P \in m \vee Q \in n$
(7)	$(P,m),(n,n)$	–	$P \in m$

(O2) Given two distinct points P and Q, fold $\mathcal{O}$ along the unique line to superpose P and Q.

(O3) Given two distinct lines m and n, fold $\mathcal{O}$ along a line to superpose m and n.

(O4) Given a line m and a point P, fold $\mathcal{O}$ along the unique line passing through P to superpose m onto itself.

(O5) Given a line m, a point P not on m and a point Q, fold $\mathcal{O}$ along a line passing through Q to superpose P and m.

(O6) Given two lines m and n, a point P not on m and a point Q not on n, where m and n are distinct or P and Q are distinct, fold $\mathcal{O}$ along a line to superpose P and m, and Q and n.

(O7) Given two lines m and n and a point P not on m, fold $\mathcal{O}$ along the unique line to superpose P and m, and n onto itself.

The above statements of (O1) $\sim$ (O7) include the conditions that eliminates degeneracy and incidence. These conditions correspond to the negations of the propositions of third and fourth column of Table 2 in natural language.

Using ζ, we define origami constructible objects.

Definition 1 (Origami constructible objects). *Given a set of initial objects* $\mathcal{S}$ ($\subseteq \Pi \cup \mathcal{L}$), *the set of origami constructible objects is inductively defined as the least set containing origami constructible objects given in 1. $\sim$ 4.:*

1. *A point P is origami constructible, if $P \in \mathcal{S}$.*
2. *An s-pair (α, β) is origami constructible if α and β are origami constructible.*
3. *A line γ is origami constructible if $\gamma \in \zeta(\underline{s})$ and $\underline{s}$ is a sequence of origami constructible s-pairs.*
4. *The intersection of lines m and n is origami constructible if m and n are origami constructible.*

One may wonder why the reflection of an origami constructible point across an origami constructible line is not included in this definition. In fact, reflections are constructible using the operations of **H** [10]. In practice, however, reflections are treated as if they were in the above inductive definition.

5.2 Properties of Operations in H

We now study the properties of the operations in **H**. For each operation, we will show the finiteness of the number of the constructible fold lines under certain conditions. This result is important to ensure the decidability of the fold since otherwise we would have an infinite computation. For each of (O1), (O2) and (O4), we have a unique fold line.

Since all the objects that we study now are origami constructible, the sets of points and lines are now denoted by $\Pi_{\mathcal{O}}$ and $\mathcal{L}_{\mathcal{O}}$ (each subscripted by $\mathcal{O}$) in all the propositions to follow. The first two are easy ones.

Proposition 1 (Fold line of (O1))

$$\forall\, P, Q \in \Pi_{\mathcal{O}} \text{ such that } P \neq Q, \ \exists\, !\, \gamma \in \mathcal{I}(P) \cap \mathcal{I}(Q).$$

This unique γ is denoted by PQ.

Proposition 2 (Fold line of (O2))

$$\forall\, P, Q \in \Pi_{\mathcal{O}} \text{ such that } P \neq Q, \ \exists\, !\, \gamma = P\updownarrow Q.$$

Proposition 3 (Fold line of (O4))

$$\forall\, m \in \mathcal{L}_{\mathcal{O}} \, \forall\, P \in \Pi_{\mathcal{O}} \ \exists\, !\ \gamma \in \mathcal{B}(m) \cap \mathcal{I}(P)$$

A fold line in (O5) is determined by s-pairs (P, m) and (Q, Q) under the condition of $P \notin m$. The fold in (O5) is impossible in certain configurations. The following proposition more sharply describes this property.

Proposition 4 (Fold lines of (O5))

$$\forall\, m \in \mathcal{L}_{\mathcal{O}} \, \forall\, P, Q \in \Pi_{\mathcal{O}} \text{ such that } P \notin m$$

$$|\, \Gamma(P, m) \cap \mathcal{I}(Q) \,| \leqslant 2.$$

Proof. The algebraic proof of this proposition is straightforward and extendable to the general cases of conic sections. Recall that $\Gamma(P, m)$ defines the set of the tangents of the parabolas whose focus and directrix are P and m, respectively. A general form of an equation of the parabola is given by the following irreducible polynomial equation

$$f(x, y) := Ax^2 + Bxy + Cy^2 + Dx + Ey + F = 0, \tag{5.1}$$

where A, B, C, D, E and F are constants, at least one of A and B is not 0, and $B^2 = 4AC$. The tangent to the curve $f(x, y)$ at the point (X, Y) is given by

$$g(x, y) := \frac{\partial f}{\partial x}(X, Y) \cdot (x - X) + \frac{\partial f}{\partial y}(X, Y) \cdot (y - Y) = 0. \tag{5.2}$$

Let Q be (u, v). As the line g passes through Q, we have $g(u, v) = 0$. We will solve for X and Y of the system of equations

$$\{f(X, Y) = 0, \ g(u, v) = 0\}.$$

Since $g(u, v)$ is linear in X and Y, finding the solutions is reduced to solving, in X (or in Y), the (at most) second degree polynomial equation obtained from $f(X, Y) = 0$ by eliminating either Y or X. Obviously, the number of real solutions is less or equal to 2. □

Concerning (O7), the following proposition holds.

Proposition 5 (Fold lines of (O7))

$$\forall \, m, n \in \mathcal{L}_O \ \forall \, P \in \Pi_O \ \text{such that} \ P \notin m,$$

$$|\, \Gamma(P, m) \cap \mathcal{B}(n) \, | \leqslant 1.$$

Proof. The proof is similar to the proof of Proposition 4. We use the formula (5.1) there. Instead of the condition that the tangent passes through a particular point, we impose the condition that the slope of the tangent at point (X, Y) is given, say $k(\neq \infty)$, in this proposition. From Eq. (5.1), we have the equation representing the tangent at (X, Y).

$$h(x, y) := D + 2Ax + By + (E + Bx + 2Cy)\frac{dy}{dx}(X, Y) = 0 \qquad (5.3)$$

Since $k(= \frac{dy}{dx}(X, Y))$ is given, all we need is to solve for X and Y in the system of equations

$$\{f(X, Y) = 0, \ D + 2AX + BY + (E + BX + 2CY)k = 0\} \qquad (5.4)$$

It is easy to see that we have at most two real solutions for the pair (X, Y).

However, when $B^2 = 4AC$, which is the case of the parabola, we have at most one real solution obtained by an easy symbolic computation using computer algebra system. □

Most interesting case is (O6), which actually gives extra power over SEC.

Proposition 6 (Fold lines of (O6))

$$\forall \, m, n \in \mathcal{L}_O \ \forall \, P, Q \in \Pi_O \ \text{such that} \ P \notin m \wedge Q \notin n$$
$$\neg(P = Q \wedge m = n) \Rightarrow$$
$$\text{if} \ m \parallel n \ \text{then} \ |\, \Gamma(P, m) \cap \Gamma(Q, n) \, | \leqslant 2$$
$$\text{else} \ 1 \leqslant |\, \Gamma(P, m) \cap \Gamma(Q, n) \, | \leqslant 3. \qquad (5.5)$$

Proof. Instead of the general equation (5.1) of the conic section, we use the following equation for the parabola defined by the focus (u, v) and the directrix $ax + by + c = 0$.

$$f(x, y) := (a^2 + b^2) ((x - u)^2 + (y - v)^2) - (ax + by + c)^2 = 0. \qquad (5.6)$$

We only have to consider the cases of $m \neq n$ and of $P \neq Q \wedge m = n$. We consider the former, first. Let $f_i(x, y)$ be the function given in (5.6) with all the constants a, b, c, u and v being indexed by i.

Let P and Q be points at (u_1, v_1) and at (u_2, v_2) respectively, and m and n be the line $a_1 x + b_1 y + c_1 = 0$, and $a_2 x + b_2 y + c_2 = 0$, respectively. Note that we can give a unique representation for the same line, so that the two lines are equal iff each coefficient a, b and c for each equation are equal. Now, let f_1 and f_2 be the parabolas defined by P and m, and by Q and n, respectively.

We distinguish the following two cases.

1. $m \nparallel n$
   As in the proof of Proposition 5, we derive the tangent h_1 with the slope t at point (X_1, Y_1) on $f_1(x, y) = 0$, and the tangent h_2 with slope t at point (X_2, Y_2) on $f_2(x, y) = 0$. The system $\{f_1(X_1, Y_1) = 0, h_1(X_1, Y_1) = 0\}$ yields X_1 and Y_1 as functions of t. Similarly, we obtain X_2 and Y_2 as functions of t. Since $(Y_1 - Y_2) - t(X_1 - X_2) = 0$, we have the polynomial equation, whose polynomial is degree 3 in t. Hence, the number of distinct real solutions is 1, 2 or 3.

2. $m \parallel n$
   Similarly to case 1., we obtain the polynomial equation of degree 2 in t. Hence we have 1 or 2 distinct real solutions.

What remains to be considered is the case of $P \neq Q \wedge m = n$. Similarly to the case 2, above, we obtain the polynomial equation of degree 2 in t. Furthermore, the discriminant of the obtained equation is easily shown to be non-negative. Hence, the relation (5.5) follows.   □

Operation (O3) is a special case of (O6) with $m = n$ and $P \neq Q$. In this case, the fold operation is about superposing the two lines PQ and m. As the corollary of Proposition 6.

Proposition 7 (Fold lines of (O3))

$$\forall\, m \in \mathcal{L}_\mathcal{O} \; \forall\, P, Q \in \Pi_\mathcal{O} \;\; P \neq Q \Rightarrow$$
$$1 \leqslant |\, \Gamma(P, m) \cap \Gamma(Q, m)\, | \leqslant 2. \qquad (5.7)$$

6   General Origami Principle

Since the algebraic interpretation of (O6) can be expressed by a cubic equation, a natural question is whether (O6) can do all the rest of fold operations of **H** with certain side conditions. The answer is basically, *yes*, but we need to carefully analyze the degenerate and incident cases, which will form the premise of the implicational formula of Lemma 1 that we will prove next.

We start with the general origami principle, which we denote by **G**, that consists of the following single operation.

(G) Given two points P and Q and two lines m and n, fold $\mathcal{O}$ along a line to superpose P and m, and Q and n.

Operation (G) is obtained by removing all the side conditions of (O6). Martin's book [10] defines a fundamental fold operation that is operation (G) with the following finiteness condition: "If for two given points P and Q and for given lines p and q there are only a finite number of lines t such that both P^t is on p and Q^t is on q". Martin uses the notation P^t to denote the reflection of point P. He further showed that some simpler operations (Yates postulates) can be derived from the fundamental fold operation. We extend this by showing that all Huzita's fold operations can be a achieved using (G), in particular under what conditions a finite number of fold lines is achieved. We refine the above Martin's statement using the results obtained so far in this paper.

We will show how the degenerate and incident cases of (G) realize the rest of the operations. We first consider the degenerate case of (G), i.e. $m = n \wedge P = Q$. This case generates the infinite set of fold lines $\Gamma(P, m)$. Furthermore, when the arguments of (O6) are more constrained, (O6) is reduced to (O2) and (O3). Suppose (O6) is given two s-pairs (P, m) and (Q, n), and further that $P \in n \wedge Q \in m$, we have Lemmas 1 and 2 below. In the following, we denote by $\{Oi\}$, $i = 1, \ldots, 7$, the set of fold lines that operation (Oi) can generate.

Lemma 1. $\forall$ s-pairs (P, m) and (Q, n) that are origami constructible, if $m \neq n \wedge P = Q \wedge (P \in n \wedge Q \in m)$ then $\{O6\} \subseteq \{O3\}$.

Proof. (Sketch) To perform (O6), P and Q have to be the intersection of m and n. (O6) then generates the two bisectors of the angle formed by m and n. Those lines are constructible by (O3) using m and n.   □

Lemma 2. $\forall$ s-pairs (P, m) and (Q, n) that are origami constructible and satisfy $(P \notin m \wedge Q \notin n)$, if $m \neq n \wedge P \neq Q \wedge (P \in n \wedge Q \in m)$, then $\{O6\} \subseteq \{O2\} \cup \{O3\}$.

Proof. (Sketch) Under the condition $m \neq n \wedge P \neq Q \wedge (P \in n \wedge Q \in m)$, (O6) generates three fold lines, i.e. $P \updownarrow Q$ and the two bisectors of the angle formed by m and n. The first one is constructible by (O2) (cf. Fig. 1(a)), and the latter ones by (O3) (cf. Fig. 1(b)).   □

Theorem 1. $\forall$ s-pairs (P, m) and (Q, n) that are origami constructible,

$$\neg((P \in m \wedge Q \in n \wedge (m \parallel n \vee P = Q)) \vee$$
$$(P \notin m \wedge Q \notin n \wedge m = n \wedge P = Q)) \Rightarrow$$

$$\{G\} = \bigcup_{i=1,\ldots,7} \{Oi\}.$$

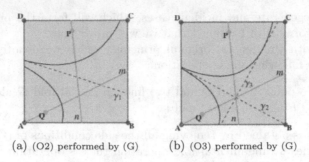

(a) (O2) performed by (G)   (b) (O3) performed by (G)

Fig. 1. (O2) and (O3) by (G) when $m \neq n \wedge P \neq Q$

Proof. We first prove that (G) is reduced to (O1), (O4), (O5), (O6) and (O7) under certain configurations of the parameters. This implies that under such conditions, $\{G\} \subseteq \{Oi\}$, where $i = 1, 4, 5, 6$ and 7.

We distinguish four cases.

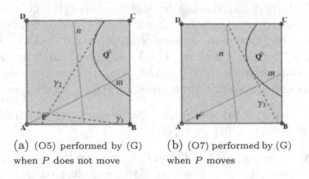

(a) (O5) performed by (G)   (b) (O7) performed by (G)
when P does not move   when P moves

Fig. 2. Incident case of (G) for $P \in m \wedge Q \notin n$

1. $P \notin m \wedge Q \notin n$
   If $(m = n \wedge P = Q)$, i.e. the degenerate case, then (G) is an undefined operation since it generates the infinite set $\Gamma(P, m)$. Otherwise, (G) performs (O6).
2. $P \in m \wedge Q \notin n$
   We further distinguish two cases of fold; P does not move, and P moves along m. In the former case, the fold line passes through P and superposes Q and n, which is the case of (O5) as shown in Fig. 2(a). In the latter case, the fold line is a perpendicular to m and superposes Q and n, which is the case of (O7) (cf. Fig. 2(b)).
3. $P \notin m \wedge Q \in n$
   Similarly to the case 1, we have the cases of (O5) and (O7).

4. $P \in m \wedge Q \in n$

We further distinguish the following four cases:

(a) $m \parallel n$

The fold lines are perpendiculars common to the lines m and n. They are infinite and form the set $\mathcal{B}(m)$.

(b) $\neg(m \parallel n) \wedge (P = Q)$

This is the case that P is the intersection of m and n. Any line passing through P is the fold line. Therefore, we have the set of infinite number of fold lines $\mathcal{I}(P)$. In this case neither P or Q does move by the fold.

(c) $\neg(m \parallel n) \wedge (P \neq Q)$

We distinguish the following three cases:

i. P moves and Q does not move

The fold line is the perpendicular to m passing through Q that superposes P and m. This is the case of (O4) (cf. Fig. 3(a)).

ii. Q moves and P does not move.

Similarly to the above case, we have the case of (O4).

iii. Neither P or Q moves.

The fold line is PQ constructible by (O1) (cf. Fig. 3(b)).

Table 3 summarizes the relations of (G) and corresponding operations of **H** for all possible combinations of the conditions.

The condition to eliminate the infinite cases are as follows.

$$(P \in m \wedge Q \in n \wedge (m \parallel n \vee P = Q))\vee$$
$$(P \notin m \wedge Q \notin n \wedge m = n \wedge P = Q)$$

Furthermore, by Lemmas 1 and 2, (O6) can be reduced to (O2) and (O3) under certain conditions. Therefore we obtain the following result

$$\neg((P \in m \wedge Q \in n \wedge (m \parallel n \vee P = Q))\vee$$
$$(P \notin m \wedge Q \notin n \wedge m = n \wedge P = Q)) \Rightarrow$$
$$\{G\} \subseteq \bigcup_{i=1,\ldots,7} \{Oi\}.$$

The relation

$$\{G\} \supseteq \bigcup_{i=1,\ldots,7} \{Oi\}$$

can be shown as follows. For each (Oi) we add parameters that satisfy the constraints to (O6) operation as shown in Table 3 in the case of (O1), (O4) - (O7) and the conditions stated in Lemmas 1 and 2 in the case of (O2) and (O3). □

Theorem 1 states that the principle **G** is as good as **H**, although **G** is much simpler *under the condition*

$$\neg((P \in m \wedge Q \in n \wedge (m \parallel n \vee P = Q))\vee$$
$$(P \notin m \wedge Q \notin n \wedge m = n \wedge P = Q)).$$

Table 3. (G) to perform (O1), (O4) - (O7)

incidence	degeneracy	operation	movement
$P \in m, Q \in n$	$m \parallel n$	$\mathcal{B}(m)$	$(\leftrightarrow, \leftrightarrow)$
	$\neg(m \parallel n) \wedge P = Q$	$\mathcal{I}(P)$	$(\cdot, \cdot)$
	$\neg(m \parallel n) \wedge P \neq Q$	(O1)	$(\cdot, \cdot)$
	$\neg(m \parallel n) \wedge P \neq Q$	(O4)	$(\leftrightarrow, \cdot)$
	$\neg(m \parallel n) \wedge P \neq Q$	(O4)	$(\cdot, \leftrightarrow)$
$P \in m, Q \notin n$		(O5)	$(\cdot, *)$
		(O7)	$(\leftrightarrow, *)$
$P \notin m, Q \in n$		(O5)	$(*, \cdot)$
		(O7)	$(*, \leftrightarrow)$
$P \notin m, Q \notin n$	$m = n \wedge P = Q$	$\Gamma(P, m)$	$(*, *)$
	$\neg(m = n \wedge P = Q)$	(O6)	$(*, *)$

Note:

- Expression (x, y) denotes the movements x and y of points P and Q, respectively.
- We denote movement x (or y) by symbols: "move" by "$\leftrightarrow$", "non-move" by "$\cdot$" and "do-not-care" by "$*$".

So let us define $\mathbf{G}'$ as $\mathbf{G}$ with the above condition. Nevertheless, $\mathbf{G}'$ has the following drawback. $\mathbf{G}'$ may create lines whose geometrical properties are different. During origami construction, a fold step may give rise to multiple possible fold lines. The user should choose a suitable fold line among the possible ones. However, in proving geometrical properties by algebraic methods like Gröbner bases, this is likely to cause problems, since the property that we want to prove may be true only for certain choices. For example, when $P \in m$ and $Q \in n$, $\mathbf{G}'$ generates two kinds of fold lines whose geometrical meaning are different, namely those by (O4) and (O1). In Fig. 3(a), the fold line γ_1 is perpendicular to m, whereas in Fig. 3(b), γ_2 is not necessary perpendicular to m. Although, the user chooses either γ_1 or γ_2 to perform the construction, the proof by Gröbner bases includes both cases. If the property that we want to prove depends on the perpendicularity of the fold line and line m, then the proof fails since perpendicularity doesn't hold for γ_2.

7   Fold with Conic Sections

We further explore the possibility of strengthening the power of origami. We extend Huzita's basic operations to allow solving polynomial equations of certain degrees while maintaining the manipulability of origami by hand. It has been shown in [9] that an extension that combines the use of compass with origami leads to interesting origami constructions, but does not increase the construction power of origami beyond what is constructible by $\mathbf{H}$. The extension generates polynomial equations of degree 4, which can be reduced to equations of degree 3.

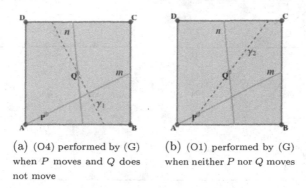

(a) (O4) performed by (G) when P moves and Q does not move

(b) (O1) performed by (G) when neither P nor Q moves

Fig. 3. Incident case of (G) for $P \in m \wedge Q \in n$

It is also possible to increase origami power by allowing multi-fold as suggested by Alperin and Lang [2]. Although the m-fold method generates an arbitrarily high degree polynomial, accurately folding origami by m-lines simultaneously would be difficult even for $m = 2$.

We further explore the foldability involving superposition of points and more general curves, which are still constructible using simple tools. In this section, we study the superposition with conic sections and describe the algebraic properties of the fold operation that superposes one point and a line, and superposes another point and a conic section assumed to be on the origami. This operation is realizable by hand and furthermore we expect to have a finite number of fold lines, which ensures the foldability. We consider a fold operation that simultaneously superposes two points with two conic sections to be difficult to perform by hand. Besides, folding to superpose a point with a conic section, with other combinations of simultaneous superpositions involving points and lines can be reduced to a more general one: superposition of two points with a line and a conic section.

To illustrate folding with conic sections by hand, an ellipse, parabola and hyperbola can be drawn on origami using pins, strings, a pencil and a straightedge, where only origami constructible points and lengths are used. Abstracting from the method used to draw a particular conic section on origami, we state the following fold operation in general:

– Given two points P and Q, a line m and a conic section $\mathcal{C}$, where P is not on $\mathcal{C}$ and Q is not on m, fold $\mathcal{O}$ along a line to superpose P and m, and Q and $\mathcal{C}$.

With little modification of the analysis performed with (O6) in Section 5.2, we obtain the following result, which corresponds to Proposition 6 for (O6).

Proposition 8. *Given origami constructible points P at (a, b) and Q at (c, d), an origami constructible line $m := y = 0$, and a conic section $\mathcal{C} := Ax^2 + Bxy + Cy^2 + Dx + Ey + F = 0$, where coefficients A, B, ..., F are origami constructible numbers and not all A, B and C are zero. We assume that P is not on m and*

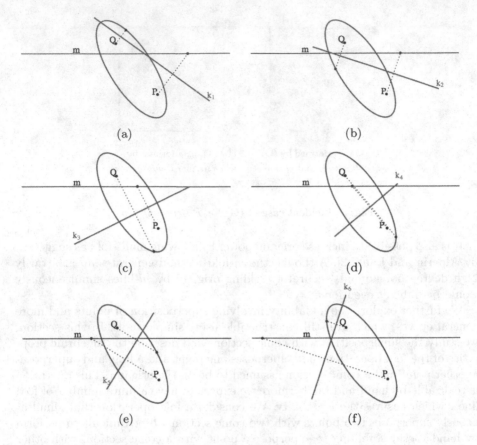

Fig. 4. Fold lines $k_1, \cdots, k_6$ whose slopes are the six distinct real solutions of the equation $16t^6 - 78t^5 + 84t^4 + 39t^3 - 66t^2 + t + 8 = 0$

Q is not on C. Let γ be the fold line to superpose P and m, and Q and C. Then the slope t of γ satisfies the following polynomial equation of degree six in t.

$$bBc + Ac^2 + b^2C - Bcd - 2bCd + Cd^2 + cD + bE - dE + F +$$
$$(-b^2B - 2Abc - 2aBc + 2Bc^2 - 4abC + 4bcC + 3bBd +$$
$$4Acd + 4aCd - 4cCd - 2Bd^2 - bD + 2dD - 2aE + 2cE)\, t +$$
$$(Ab^2 + 4abB + 4aAc - 4bBc - 2Ac^2 + 4a^2C - 2b^2C -$$
$$8acC + 4c^2C - 4Abd - 6aBd + 6Bcd + 4bCd + 4Ad^2 -$$
$$2Cd^2 + 2aD + 2F)\, t^2 +$$
$$(-4aAb - 4a^2B + 2b^2B + 4Abc + 6aBc - 2Bc^2 + 4abC -$$
$$4bcC + 8aAd - 4bBd - 4Acd - 4aCd + 4cCd + 2Bd^2 +$$
$$2dD - 2aE + 2cE)\, t^3 +$$
$$(4a^2A - 2Ab^2 - 4abB - 4aAc + 3bBc + Ac^2 + b^2C + Bcd -$$
$$4Abd + 2aBd - 2bCd + Cd^2 + 2aD - cD - bE + dE + F)\, t^4 +$$
$$(4aAb - b^2B - 2Abc + bBd + bD)\, t^5 + Ab^2\, t^6$$

Proof. (Sketch) Let points U and V be the reflections of P and Q respectively across the fold line γ. Point U is on line m and point V is on the given conic section $\mathcal{C}$. Furthermore, fold line γ is the perpendicular bisector of segments PU and QV. From the equations of these relations, with algebraic manipulation by a computer algebra system, we can derive the above degree six equation in slope t of line γ. $\square$

This equation looks laborious; one should only note that it is an equation in t of degree six over the field of origami constructible numbers.

In the example shown in Fig. 4, we assume origami constructible points P at $(3, -4)$, Q at $(-1, 1)$ and $m := y = 0$. The conic section $\mathcal{C} := 2x^2 + 2xy + y^2 + x + 2y - 10 = 0$ is an ellipse depicted in Fig. 4. Giving concrete numerical values that realize the figure, we obtain the following equation for t.

$$16t^6 - 78t^5 + 84t^4 + 39t^3 - 66t^2 + t + 8 = 0 \qquad (7.1)$$

Solving Eq. (7.1) using computer algebra system yields six real solutions that correspond to six possible fold lines $k_1, \cdots, k_6$ in Fig. 4. The same operation can be performed by hand, obtaining fold lines with certain slopes. Each slope value is one real solution to the equation.

8   Concluding Remarks

We reformulated the Huzita's operations giving them precise definitions with side conditions that eliminate the degenerate and incident cases. We showed that for each of the reformulated operations only a finite number of fold lines is possible. We gave an alternative single operation based on operation (O6) and showed how each of the reformulated operations can be performed using the new one. Furthermore, we investigated the combination of origami operations and conic sections. We showed that finding a fold line that superposes two points, one with a line and the other with a conic section, is reduced to solving an equation of degree six. We can think of two directions for future work of this research.

First, the principles of folding presented in this paper have been worked out carefully, so that they can be formalized in a proof assistant. Starting from a formalization of the basic geometric concepts, one can formally define the lines (or sets of lines) that arise from particular fold operations. This can be used to specify the superpositions that arise from the composition of fold operations, and the set of origami constructible points and lines. We imagine that such a development would give a basis for a formalized origami theory. Recently, it has been shown [8], that the decision procedures already present in modern proof assistants combined with the symbolic computation procedures are strong enough to solve many of the goals arising in computational origami problems.

Second, further investigation of fold operations involving conic sections is required to give exact definitions of the fold operations with their degenerate and incident cases. We showed that superposing two points onto line and conic section gives rise to equation of degree six. However the bigger question is whether this fold operation would solve all the equations of degree five and six. In other words, can we find an algorithm for translating degree five and degree six equations, possibly with certain conditions, into origami fold problems?

References

1. Alperin, R.C.: A Mathematical Theory of Origami Constructions and Numbers. New York Journal of Mathematics 6, 119–133 (2000)
2. Alperin, R.C., Lang, R.J.: One-, Two, and Multi-fold Origami Axioms. In: Origami^4 Fourth International Meeting of Origami Science, Mathematics and Education (4OSME), pp. 371–393. A K Peters Ltd. (2009)
3. Ghourabi, F., Ida, T., Takahashi, H., Kasem, A.: Reasoning Tool for Mathematical Origami Construction. In: CD Proceedings of the International Symposium on Symbolic and Algebraic Computation, ISSAC 2009 (2009)
4. Huzita, H.: Axiomatic Development of Origami Geometry. In: Proceedings of the First International Meeting of Origami Science and Technology, pp. 143–158 (1989)
5. Ida, T., Kasem, A., Ghourabi, F., Takahashi, H.: Morley's theorem revisited: Origami construction and automated proof. Journal of Symbolic Computation 46(5), 571–583 (2011)
6. Jones, A., Morris, S.A., Pearson, K.R.: Abstract Algebra and Famous Impossibilities. Springer-Verlag New York, Inc. (1991)
7. Justin, J.: Résolution par le pliage de l'équation du troisième degré et applications géométriques. In: Proceedings of the First International Meeting of Origami Science and Technology, pp. 251–261 (1989)
8. Kaliszyk, C., Ida, T.: Proof Assistant Decision Procedures for Formalizing Origami. In: Davenport, J.H., Farmer, W.M., Urban, J., Rabe, F. (eds.) Calculemus/MKM 2011. LNCS, vol. 6824, pp. 45–57. Springer, Heidelberg (2011)
9. Kasem, A., Ghourabi, F., Ida, T.: Origami Axioms and Circle Extension. In: Proceedings of the 26th Symposium on Applied Computing (SAC 2011), pp. 1106–1111. ACM Press (2011)
10. Martin, G.E.: Geometric Constructions. Springer-Verlag New York, Inc. (1998)
11. Wantzel, P.L.: Recherches sur les moyens de connaître si un problème de géométrie peut se résoudre avec la règle et le compas. Journal de Mathématiques Pures et Appliquées, 366–372 (1984)

On the Formal Analysis of Geometrical Optics in HOL

Umair Siddique, Vincent Aravantinos, and Sofiène Tahar

Department of Electrical and Computer Engineering,
Concordia University, Montreal, Canada
{muh_sidd,vincent,tahar}@ece.concordia.ca

Abstract. Geometrical optics, in which light is characterized as rays, provides an efficient and scalable formalism for the modeling and analysis of optical and laser systems. The main applications of geometrical optics are in stability analysis of optical resonators, laser mode locking and micro opto-electro-mechanical systems. Traditionally, the analysis of such applications has been carried out by informal techniques like paper-and-pencil proof methods, simulation and computer algebra systems. These traditional techniques cannot provide accurate results and thus cannot be recommended for safety-critical applications, such as corneal surgery, process industry and inertial confinement fusion. On the other hand, higher-order logic theorem proving does not exhibit the above limitations, thus we propose a higher-order logic formalization of geometrical optics. Our formalization is mainly based on existing theories of multivariate analysis in the HOL Light theorem prover. In order to demonstrate the practical effectiveness of our formalization, we present the modeling and stability analysis of some optical resonators in HOL Light.

1 Introduction

Different characterizations of light lead to different fields of optics such as quantum optics, electromagnetic optics, wave optics and geometrical optics. The latter describes light as rays which obey geometrical rules. The theory of geometrical optics can be applied for the modeling and analysis of physical objects with dimensions greater than the wavelength of light. Geometrical optics is based on a set of postulates which are used to derive the rules for the propagation of light through an optical medium. These postulates can be summed up as follows: light travels in the form of rays emitted by a source; an optical medium is characterized by its refractive index; light rays follow Fermat's principle of least time [19].

Optical components, such as thin lenses, thick lenses and prisms are usually centered about an optical axis, around which rays travel at small inclinations (angle with the optical axis). Such rays are called *paraxial rays* and this assumption provides the basis of *paraxial optics* which is the simplest framework of geometrical optics. The paraxial approximation explains how light propagates

T. Ida and J. Fleuriot (Eds.): ADG 2012, LNAI 7993, pp. 161–180, 2013.
© Springer-Verlag Berlin Heidelberg 2013

through a series of optical components and provides diffraction-free descriptions of complex optical systems. The change in the position and inclination of a paraxial ray as it travels through an optical system can be efficiently described by the use of matrices [11]. This matrix formalism (called *ray-transfer matrices*) of geometrical optics provides accurate, scalable and systematic analysis of real-world complex optical and laser systems. This fact has led to the widespread usage of ray-transfer matrices in the modeling and analysis of critical physical systems. Typical applications of ray-transfer matrices include analysis of a laser beam propagation through some optical setup [11], the stability analysis of laser or optical resonators [13], laser mode-locking, optical pulse transmission [15] and analysis of micro opto-electro-mechanical systems (MOEMS) [28]. Another promising feature of the matrix formalism of geometrical optics is the prediction of design parameters for physical experiments, e.g., recent dispersion-managed soliton transmission experiment [14] and invention of the first single-cell biological lasers [5].

Traditionally, the analysis of geometrical optics based models has been done using paper-and-pencil proof methods [11,15,14]. However, considering the complexity of present age optical and laser systems, such an analysis is very difficult if not impossible, and thus quite error-prone. Many examples of erroneous paper-and-pencil based proofs are available in the open literature, a recent one can be found in [4] and its identification and correction is reported in [16]. One of the most commonly used computer-based analysis techniques for geometrical optics based models is numerical computation of complex ray-transfer matrices [25,12]. Optical and laser systems involve complex and vector analysis and thus numerical computations cannot provide perfectly accurate results due to the inherent incomplete nature of the underlying numerical algorithms. Another alternative is computer algebra systems [17], which are very efficient for computing mathematical solutions symbolically, but are not 100% reliable due to their inability to deal with side conditions [7]. Another source of inaccuracy in computer algebra systems is the presence of unverified huge symbolic manipulation algorithms in their core, which are quite likely to contain bugs. Thus, these traditional techniques should not be relied upon for the analysis of critical laser and optical systems (e.g., corneal surgery [27]), where inaccuracies in the analysis may even result in the loss of human lives.

In the past few years, higher-order logic theorem proving has been successfully used for the precise analysis of a few continuous physical systems [22,10]. Developing a higher-order logic model for a physical system and analyzing this model formally is a very challenging task since it requires both a good mathematical and physical knowledge. However, it provides an effective way for identifying critical design errors that are often ignored by traditional analysis techniques like simulation and computer algebra systems. We believe that higher-order logic theorem proving [6] offers a promising solution for conducting formal analysis of such critical optical and laser systems. Most of the classical mathematical theories behind geometrical optics, such as Euclidean spaces, multivariate analysis and complex numbers, have been formalized in the HOL Light theorem prover

[8,9]. To the best of our knowledge, the reported formalization of geometrical optics is the first of its kind. Our HOL Light developments of geometrical optics and applications presented in this paper are available for download [20] and thus can be used by other researchers and optics engineers working in industry to conduct the formal analysis of their optical systems. This paper is an extended and improved version of [21].

The rest of the paper is organized as follows: Section 2 describes some fundamentals of geometrical optics, and its commonly used ray-transfer-matrix formalism. Section 3 presents our HOL Light formalization of geometrical optics. In order to demonstrate the practical effectiveness and the use of our work, we present in Section 4 the analysis of two real-world optical resonators: Fabry-Pérot resonator and Z-shaped resonator. Finally, Section 5 concludes the paper and highlights some future directions.

2   Geometrical Optics

When a ray passes through optical components, it undergoes *translation* or *refraction*. When it comes to translation, the ray simply travels in a straight line from one component to the next and we only need to know the thickness of the translation. On the other hand, refraction takes place at the boundary of two regions with different refractive indices and the ray follows the law of refraction, i.e., the angle of refraction relates to the angle of incidence by the relation $n_0 \sin(\phi_0) = n_1 \sin(\phi_1)$, called *Snell's law* [19], where n_0, n_1 are the refractive indices of both regions and ϕ_0, ϕ_1 are the angles of the incident and refracted rays, respectively, with the normal to the surface. In order to model refraction, we thus need the normal to the refracting surface and the refractive indices of both regions.

In order to introduce the matrix formalism of geometrical optics, we consider the propagation of a ray through a spherical interface with radius of curvature R between two mediums of refractive indices n_0 and n_1, as shown in Figure 1. Our goal is to express the relation between the incident and refracted rays. The trajectory of a ray as it passes through various optical components can be specified by two parameters: its distance and angle with the optical axis. Here, the distances with respect to the optical axis of the incident and refracted rays are r_1 and r_0, respectively. Since the thickness of the surface is assumed to be very small, we consider that $r_1 = r_0$. Here, ϕ_0 and ϕ_1 are the angles of the incident and refracted rays with the normal to the spherical surface, respectively. On the other hand, θ_0 and θ_1 are the angles of the incident and refracted rays with the optical axis.

Applying Snell's law at the interface, we have $n_0 \sin(\phi_0) = n_1 \sin(\phi_1)$, which, in the context of the paraxial approximation, reduces to the form $n_0 \phi_0 = n_1 \phi_1$ since $\sin(\phi) \simeq \phi$ if ϕ is small. We also have $\theta_0 = \phi_0 - \psi$ and $\theta_1 = \phi_1 - \psi$, where ψ is the angle between the surface normal and the optical axis. Since $\sin(\psi) = \frac{r_0}{R}$, then $\psi = \frac{r_0}{R}$ by the paraxial approximation again. We can deduce that:

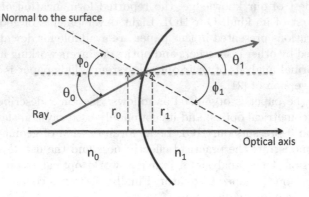

Fig. 1. Spherical interface

$$\theta_1 = \left(\frac{n_0 - n_1}{n_1 R}\right) r_0 + \left(\frac{n_0}{n_1}\right) \theta_0 \tag{1}$$

So, for a spherical surface, we can relate the refracted ray with the incident ray by a matrix relation using equation (1) as follows:

$$\begin{bmatrix} r_1 \\ \theta_1 \end{bmatrix} = \begin{bmatrix} 1 & 0 \\ \frac{n_0 - n_1}{n_1 R} & \frac{n_0}{n_1} \end{bmatrix} \begin{bmatrix} r_0 \\ \theta_0 \end{bmatrix}$$

Thus the propagation of a ray through a spherical interface can be described by a 2×2 matrix generally called, in the literature, *ABCD matrix*. It is actually possible to obtain such a 2×2 matrix relating r_1, θ_1 and r_0, θ_0 for many optical components [19].

If we have an optical system consisting of k optical components, then we can trace the input ray R_i through all optical components using multiplication of the matrices of all optical components as follows:

$$R_o = (M_k.M_{k-1}....M_1).R_i \tag{2}$$

where R_o is the output ray and R_i is the input ray.

3 Formalization of Geometrical Optics

In this section, we present a brief overview of our higher-order logic formalization of geometrical optics. The formalization consists of three parts: 1) fundamental concepts of optical systems structures and light ray; 2) frequently used optical components (i.e., thin lens, thick lens and plane parallel plate); 3) optical resonators and their stability.

3.1   Optical System Structure and Ray

The formalization is two-fold: first, we model the geometry and physical parameters of an optical system; second, we model the physical behavior of a ray when it goes through an optical interface. Afterwards, we derive the ray-transfer matrices of the optical components, as explained in Section 2.

An optical system is a sequence of optical components, which consists of free spaces and optical interfaces. We define interfaces by an inductive data type enumerating their different kinds and their corresponding parameters:

Definition 1 (Optical Interface and System)
```
define_type "optical_interface = plane | spherical ℝ"
define_type "interface_kind = transmitted | reflected"
new_type_abbrev ("free_space",':ℝ × ℝ')
new_type_abbrev ("optical_component",
           ':free_space × optical_interface × interface_kind')
new_type_abbrev ("optical_system",
           ':optical_component list × free_space')
```

An optical component is made of a free space, which is formalized by a pair of real numbers representing the refractive index and width of free space, and an optical interface which is of two kinds: plane or spherical, yielding the corresponding constructors as shown in Figure 2. A spherical interface takes a real number representing its radius of curvature. Finally, an optical system is a list of optical components followed by a free space. When passing through an interface, the ray is either transmitted or reflected. In this formalization, this information is also provided in the type of optical components, as shown by the use of the type `interface_kind`. Note that this datatype can easily be extended to many other optical components if needed.

A value of type `free_space` does represent a real space only if the refractive index is greater than zero. In addition, in order to have a fixed order in the representation of an optical system, we impose that the distance of an optical interface relative to the previous interface is greater or equal to zero. We also need to assert the validity of a value of type `optical_interface` by ensuring that the radius of curvature of spherical interfaces is never equal to zero. This yields the following predicates:

Definition 2 (Valid Free Space and Valid Optical Interface)
⊢ is_valid_free_space ((n,d):free_space) ⇔ 0 < n ∧ 0 ≤ d
⊢ (is_valid_interface plane ⇔ T) ∧
   (is_valid_interface (spherical R) ⇔ 0 <> R)

Then, by ensuring that this predicate holds for every component of an optical system, we can characterize valid optical systems as follows:

Definition 3 (Valid Optical Component)
⊢ ∀fs i ik. is_valid_optical_component ((fs,i,ik):optical_component)
   ⇔ is_valid_free_space fs ∧ is_valid_interface i

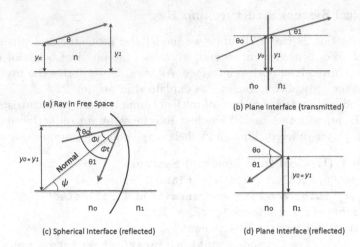

Fig. 2. Behavior of ray at different interfaces

Definition 4 (Valid Optical System)
⊢ ∀os fs. is_valid_optical_system ((cs,fs):optical_system) ⇔
    ALL is_valid_optical_component cs ∧ is_valid_free_space fs

where ALL is a HOL Light library function which checks that a predicate holds for all the elements of a list. We conclude our formalization of an optical system by defining the following helper function to retrieve the refractive index of the first free space in an optical system:

Definition 5 (Refractive Index of First Free Space)
⊢ (head_index ([],(n,d)) = n) ∧
    (head_index (CONS ((n,d),i) cs, (nt,dt)) = n)

We can now formalize the physical behavior of a ray when it passes through an optical system. We only model the points where it hits an optical interface (instead of modeling all the points constituting the ray). So it is sufficient to just provide the distance of every of these hitting points to the axis and the angle taken by the ray at these points. Consequently, we should have a list of such pairs (*distance, angle*) for every component of a system. In addition, the same information should be provided for the source of the ray. For the sake of simplicity, we define a type for a pair (*distance, angle*) as ray_at_point. This yields the following definition:

Definition 6 (Ray)
new_type_abbrev ("ray_at_point", ':ℝ × ℝ')
new_type_abbrev ("ray", ':ray_at_point × ray_at_point ×
                        (ray_at_point × ray_at_point) list')

The first ray_at_point is the pair (*distance, angle*) for the source of the ray, the second one is the one after the first free space, and the list of ray_at_point pairs represents the same information for the interfaces and free spaces at every hitting point of an optical system. Once again, we specify what is a valid ray by using some predicates. First of all, we define what is the behavior of a ray when it is traveling through a free space. This requires the position and orientation of the ray at the previous and current point of observation, and the free space itself. This is shown in Figure 2(a).

Definition 7 (Behavior of a Ray in Free Space)
$\vdash$ is_valid_ray_in_free_space (y_0,θ_0) (y_1,θ_1) $((n,d):$free_space$)$ $\Leftrightarrow$
$$y_1 = y_0 + d * \theta_0 \wedge \theta_0 = \theta_1$$

Next, we define what is the valid behavior of a ray when hitting a particular interface. This requires the position and orientation of the ray at the previous and current interface, and the refractive index before and after the component. Then the predicate is defined by case analysis on the interface and its type as follows:

Definition 8 (Behavior of a Ray at Given Interface)
$\vdash$ (is_valid_ray_at_interface (y_0,θ_0) (y_1,θ_1) n_0 n_1 plane transmitted
$\Leftrightarrow$ $y_1 = y_0 \wedge n_0 * \theta_0 = n_1 * \theta_1)$ $\wedge$
(is_valid_ray_at_interface (y_0,θ_0) $(y1,\theta_1)$ n_0 n_1 (spherical R)
transmitted $\Leftrightarrow$ let $\phi_i = \theta_0 + \frac{y_1}{R}$ and $\phi_t = \theta_1 + \frac{y_1}{R}$ in
$y_1 = y_0 \wedge n_0 * \phi_i = n_1 * \phi_t)$ $\wedge$
(is_valid_ray_at_interface (y_0,θ_0) (y_1,θ_1) n_0 n_1 plane reflected
$\Leftrightarrow$ $y_1 = y_0 \wedge n_0 * \theta_0 = n_0 * \theta_1)$ $\wedge$
(is_valid_ray_at_interface (y_0,θ_0) (y_1,θ_1) n_0 n_1 (spherical R)
reflected $\Leftrightarrow$ let $\phi_i = \frac{y_1}{R} - \theta_0$ in $y_1 = y_0 \wedge \theta_1 = -(\theta_0 + 2 * \phi_i))$

The above definition states some basic geometrical facts about the distance to the axis, and applies Snell's law to the orientation of the ray as shown in Figures 1 and 2. Note that, both to compute the distance and to apply Snell's law, we assumed the paraxial approximation in order to turn $\sin(\theta)$ into θ. Finally, we can recursively apply these predicates to all the components of a system as follows:

Definition 9 (Behavior of a Ray in an Optical System)
$\vdash$ $\forall$ sr_1 sr_2 h h' fs cs rs i ik y_0 θ_0 y_1 θ_1 y_2 θ_2 y_3 θ_3 n d n' d' .
(is_valid_ray_in_system $(sr_1,sr_2,[])$ (CONS h cs,fs) $\Leftrightarrow$ F) $\wedge$
(is_valid_ray_in_system $(sr_1,sr_2,$CONS h' rs) $([],$fs) $\Leftrightarrow$F)$\wedge$
(is_valid_ray_in_system $((y_0,\theta_0),(y_1,\theta_1),[])$ $([],$n,d) $\Leftrightarrow$
is_valid_ray_in_free_space (y_0,θ_0) (y_1,θ_1) (n,d)) $\wedge$
(is_valid_ray_in_system $((y_0,\theta_0),(y_1,\theta_1),$
CONS $((y_2,\theta_2),y_3,\theta_3)$ rs) (CONS $((n',d'),$i,ik) cs,n,d) $\Leftrightarrow$
(is_valid_ray_in_free_space (y_0,θ_0) (y_1,θ_1) (n',d') $\wedge$
is_valid_ray_at_interface (y_1,θ_1) (y_2,θ_2) n'
(head_index (cs,n,d)) i ik)) $\wedge$
(is_valid_ray_in_system $((y_2,\theta_2),(y_3,\theta_3),$rs) (cs,n,d))

The behavior of a ray going through a series of optical components is thus completely defined. Using this formalization, we verify the ray-transfer matrices as presented in Section 2. In order to facilitate formal reasoning, we define the following matrix relations for free spaces and interfaces.

Definition 10 (Free Space Matrix)

$\vdash \forall d. \ \texttt{free\_space\_matrix} \ d = \begin{bmatrix} 1 & d \\ 0 & 1 \end{bmatrix}$

Definition 11 (Interface Matrix)

$\vdash \forall n_0 \ n_1 \ R.$

$\texttt{interface\_matrix} \ n_0 \ n_1 \ \texttt{plane transmitted} = \begin{bmatrix} 1 & 0 \\ 0 & \frac{n_0}{n_1} \end{bmatrix} \ \wedge$

$\texttt{interface\_matrix} \ n_0 \ n_1 \ \texttt{(spherical R) transmitted} = \begin{bmatrix} 1 & 0 \\ \frac{n_0-n_1}{n_0*R} & \frac{n_0}{n_1} \end{bmatrix} \ \wedge$

$\texttt{interface\_matrix} \ n_0 \ n_1 \ \texttt{plane reflected} = \begin{bmatrix} 1 & 0 \\ 0 & 1 \end{bmatrix} \ \wedge$

$\texttt{interface\_matrix} \ n_0 \ n_1 \ \texttt{(spherical R) reflected} = \begin{bmatrix} 1 & 0 \\ \frac{-2}{R} & 1 \end{bmatrix}$

In the above definition, n_0 and n_1 represent the refractive indices before and after an optical interface. We use the traditional mathematical notation of matrices for the sake of clarity, but, in practice, we use the dedicated functions of HOL Light's vectors library.

Next, we verify the ray-transfer-matrix relation for free spaces:

Theorem 1 (Ray-Transfer-Matrix for Free Space)
$\vdash \forall n \ d \ y_0 \ \theta_0 \ y_1 \ \theta_1. \ \texttt{is\_valid\_free\_space} \ (n,d) \ \wedge$
$\texttt{is\_valid\_ray\_in\_free\_space} \ (y_0,\theta_0) \ (y_1,\theta_1) \ (n,d)) \implies$
$\begin{bmatrix} y_1 \\ \theta_1 \end{bmatrix} = \texttt{free\_space\_matrix} \ d \ \texttt{**} \ \begin{bmatrix} y_0 \\ \theta_0 \end{bmatrix}$

where ** represents matrix-vector or matrix-matrix multiplication. The first assumption ensures the validity of free space and the second assumption ensures the valid behavior of ray in free space. The proof of this theorem requires some properties of vectors and matrices along with some arithmetic reasoning. Next, we verify an important theorem describing the general ray-transfer-matrix relation for any interface as follows:

Theorem 2 (Ray-Transfer-Matrix any Interface)
$\vdash \forall n_0 \ n_1 \ y_0 \ \theta_0 \ y_1 \ \theta_1 \ i \ ik. \ \texttt{is\_valid\_interface} \ i \ \wedge$
$\texttt{is\_valid\_ray\_at\_interface} \ (y_0,\theta_0) \ (y_1,\theta_1) \ n_0 \ n_1 \ i \ ik \ \wedge$

$0 < n_0 \ \wedge \ 0 < n_1 \implies \begin{bmatrix} y_1 \\ \theta_1 \end{bmatrix} = \texttt{interface\_matrix} \ n_0 \ n_1 \ i \ ik \ \texttt{**} \ \begin{bmatrix} y_0 \\ \theta_0 \end{bmatrix}$

In the above theorem, both assumptions ensure the validity of the interface and behavior of ray at the interface, respectively. This theorem is easily proved by case splitting on i and ik.

Now, equipped with the above theorem, the next step is to formally verify the ray-transfer-matrix relation for a complete optical system as given in Equation 2. It is important to note that in this equation, individual matrices of optical components are composed in reverse order. We formalize this fact with the following recursive definition:

Definition 12 (System Composition)
```
⊢ system_composition ([],n,d) ⇔ free_space_matrix d ∧
  system_composition (CONS ((nt,dt),i,ik) cs,n,d) ⇔
  (system_composition (cs,n,d) **
  interface_matrix nt (head_index (cs,n,d)) i ik) **
  free_space_matrix dt
```

The general ray-transfer-matrix relation is then given by the following theorem:

Theorem 3 (Ray-Transfer-Matrix for Optical System)
```
⊢ ∀sys ray. is_valid_optical_system sys ∧
  is_valid_ray_in_system ray sys ⟹
  let (y_0,θ_0),(y_1,θ_1),rs = ray in
  let y_n,θ_n = last_ray_at_point ray in
```
$$\begin{bmatrix} y_n \\ \theta_n \end{bmatrix} = \texttt{system\_composition sys ** } \begin{bmatrix} y_0 \\ \theta_0 \end{bmatrix}$$

Here, the parameters `sys` and `ray` represent the optical system and the ray respectively. The function `last_ray_at_point` returns the last `ray_at_point` of the ray in the system. Both assumptions in the above theorem ensure the validity of the optical system and the good behavior of the ray in the system. The theorem is easily proved by induction on the length of the system and by using previous results and definitions.

This concludes our formalization of optical system structure and ray along with the verification of important properties of optical components and optical systems. The formal verification of the above theorems not only ensures the effectiveness of our formalization but also shows the correctness of our definitions related to optical systems.

3.2 Frequently Used Optical Components

In this section, we present the formal modeling and verification of the ray-transfer matrix relation of thin lens, thick lens and plane parallel plate [19], which are the most widely used components in optical and laser systems.

Generally, lenses are determined by their refractive indices and thickness. A thin lens is represented as the composition of two transmitting spherical interfaces such that any variation of ray parameters (position y and orientation θ) is neglected between both interfaces, as shown in Figure 3 (a). So, at the end, a thin lens is the composition of two spherical interfaces with a null width free space in between. We formalize thin lenses as follows:

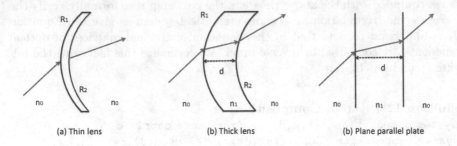

Fig. 3. Frequently used optical components

Definition 13 (Thin Lens)
⊢ ∀R₁ R₂ n₀ n₁. thin_lens R₁ R₂ n₀ n₁ =
  ([(n₀,0),spherical R₁,transmitted; (n₁,0),spherical R₂,transmitted],
  (n₀,0))

We can then prove that a thin lens is indeed a valid optical system if the corresponding parameters satisfy some constraints:

Theorem 4 (Valid Thin Lens)
⊢ ∀R₁ R₂ n₀ n₁. R₁ ≠ 0 ∧ R₂ ≠ 0 ∧ 0 < n₀ ∧ 0 < n₁ ⟹
  is_valid_optical_system (thin_lens R₁ R₂ n₀ n₁)

Now, in order to simplify the reasoning process, we define the thin lens matrix:

Definition 14 (Thin Lens Matrix)

$$⊢ ∀R_1\ R_2\ n_0\ n_1.\ \texttt{thin\_lens\_mat}\ R_1\ R_2\ n_0\ n_1 = \begin{bmatrix} 1 & 0 \\ \frac{n_1\ -\ n_0}{n_0}\left(\frac{1}{R_2} - \frac{1}{R_1}\right) & 1 \end{bmatrix}$$

Next, we verify that this matrix is indeed the ray-transfer matrix of the corresponding thin lens:

Theorem 5 (Thin Lens Matrix)
⊢ ∀R₁ R₂ n₀ n₁. R₁ ≠ 0 ∧ R₂ ≠ 0 ∧ 0 < n₀ ∧ 0 < n₁
  ⟹ system_composition (thin_lens R₁ R₂ n₀ n₁) =
  thin_lens_mat R₁ R₂ n₀ n₁

Finally, we can wrap up the behavior of a ray through a thin lens as follows, thanks to Theorem 3:

Theorem 6 (Ray-Transfer-Matrix Model of Thin Lens)
⊢ ∀R₁ R₂ n₀ n₁. R₁ ≠ 0 ∧ R₂ ≠ 0 ∧ 0 < n₀ ∧ 0 < n₁ ⟹
  (∀ray.is_valid_ray_in_system ray (thin_lens R₁ R₂ n₀ n₁)
    ⟹ (let (y₀,θ₀),(y1,theta1),rs = ray in
    (yₙ,θₙ) = last_single_ray ray in
  vector [yₙ;θₙ] = thin_lens_mat R₁ R₂ n₀ n₁ ** vector [y₀;θ₀]))

The thick lens is another useful optical component which is used in many real-world optical systems [19]. It is a composition of two spherical interfaces separated by a distance d as shown in Figure 3 (b). We formalize thick lenses as follows:

Definition 15 (Thick Lens)
⊢ ∀R_1 R_2 n_0 n_1 d. thick_lens R_1 R_2 n_0 n_1 d =
  ([(n_0,0),spherical R_1,transmitted; (n_1,d),spherical R_2,transmitted],
  (n_0,0))

Next, we verify that this lens indeed represents a valid optical system:

Theorem 7 (Valid Thick Lens)
⊢ ∀R_1 R_2 n_0 n_1. $R_1 \neq 0 \wedge R_2 \neq 0 \wedge 0 < n_0 \wedge 0 < n_1 \wedge 0 \leq d$
  ⟹ is_valid_optical_system (thick_lens R_1 R_2 n_0 n_1 d)

Again, we define the thick lens matrix:

Definition 16 (Thick Lens Matrix)
⊢ ∀R_1 R_2 n_0 n_1 d. thin_lens_mat R_1 R_2 n_0 n_1 =

$$\begin{bmatrix} 1 + \frac{d*n_0}{R_1*n_1} - \frac{1}{R1} & \frac{d*n0}{n_0} \\ -\frac{(n_0-n_1)*[d*(n_0 - n_1) + n_1*(R_1-R_2)]}{n_0*n_1*R_1*R_2} & 1 + d*\left(\frac{1}{R_2} - \frac{n_1}{n_1*R_2}\right) \end{bmatrix}$$

We verify that this matrix is indeed the ray-transfer matrix of the thick lens as follows:

Theorem 8 (Thick Lens Matrix)
⊢ ∀R_1 R_2 n_0 n_1 d. $R_1 \neq 0 \wedge R_2 \neq 0 \wedge 0 < n_0 \wedge 0 < n_1 \wedge n_1 \wedge$
  $0 \leq d$ ⟹ system_composition (thick_lens R_1 R_2 n_0 n_1) =
  thick_lens_mat R_1 R_2 n_0 n_1

We then easily obtain the ray-transfer-matrix relation for thick lenses by using Theorem 3:

Theorem 9 (Ray-Transfer-Matrix Model of Thick Lens)
⊢ ∀R_1 R_2 n_0 n_1 d. $R_1 \neq 0 \wedge R_2 \neq 0 \wedge 0 < n_0 \wedge 0 < n_1 \wedge$
  $0 \leq d$ ⟹
  (∀ ray.is_valid_ray_in_system ray (thick_lens R_1 R_2 n_0 n_1 d)
    ⟹ (let (y_0,θ_0),(y1,theta1),rs = ray in
    (y_n,θ_n) = last_single_ray ray in
  vector [y_n;θ_n] = thick_lens_mat R_1 R_2 n_0 n_1 d ** vector [y_0;θ_0]))

The plane parallel plate is another useful optical component which consists of two plane interfaces separated by some distance d as shown in Figure 3 (c). We formally model plane parallel plates as follows:

Definition 17 (Plane Parallel Plate)
⊢ ∀n_0 n_1 d. plane_parallel_plate n_0 n_1 d =
  ([(n_0,0),plane,transmitted; (n_1,d),plane,transmitted],(n_0,0))

Next, we verify this system:

Theorem 10 (Valid Plane Parallel Plate)
⊢ $\forall n_0$ n_1 d. $0 < n_0 \wedge 0 < n_1 \wedge 0 \leq d \implies$
  is_valid_optical_system (plane_parallel_plate n_0 n_1 d)

Now, we define the matrix for plane parallel plate:

Definition 18 (Plane Parallel Plate Matrix)
⊢ $\forall n_0$ n_1 d. plane_parallel_mat n_0 n_1 d = $\begin{bmatrix} 1 & d * \frac{n_0}{n_1} \\ 0 & 1 \end{bmatrix}$

Next, we verify that this matrix is indeed the ray-transfer matrix of the corresponding plane parallel plate:

Theorem 11 (Plane Parallel Matrix)
⊢ $\forall n_0$ n_1 d. $0 < n_0 \wedge 0 < n_1 \wedge 0 \leq d$
  $\implies$ system_composition (plane_parallel_plate n_0 n_1 d) =
  plane_parallel_mat n_0 n_1 d

Finally, we can verify the behavior of ray through a plane parallel plate:

Theorem 12 (Ray-Transfer-Matrix Model of Plane Parallel Plate)
⊢ $\forall n_0$ n_1 d. $0 < n_0 \wedge 0 < n_1 \wedge 0 \leq d \implies$
  ($\forall$ray.is_valid_ray_in_system ray (plane_parallel_plate n_0 n_1 d)
    $\implies$ (let (y_0, θ_0),(y1,theta1),rs = ray in
       (y_n, θ_n) = last_single_ray ray in
  vector [$y_n; \theta_n$] = plane_parallel_mat n_0 n_1 d ** vector [$y_0; \theta_0$]))

This concludes our formalization of some frequently-used components, which demonstrates how we can use our optics fundamentals formalization in order to define basic systems.

3.3   Optical Resonators and Their Stability

An optical resonator usually consists of mirrors or lenses which are configured in such a way that the beam of light confines in a closed path as shown in Figure 4. Optical resonators are fundamental building blocks of optical devices and lasers. Resonators differ by their geometry and components (interfaces and mirrors) used in their design.

Optical resonators are broadly classified as stable or unstable. Stability analysis identifies geometric constraints of the optical components which ensure that light remains inside the resonator (see Figure 5 (a)). Both stable and unstable resonators have diverse applications, e.g., stable resonators are used in the measurement of refractive index of cancer cells [24], whereas unstable resonators are used in the laser oscillators for high energy applications [23].

The stability analysis of optical resonators involves the study of infinite rays, or, equivalently, of an infinite set of finite rays. Indeed, a resonator is a closed structure terminated by two reflected interfaces and a ray reflects back and forth

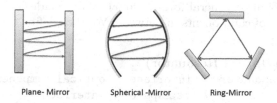

Fig. 4. Optical Resonators

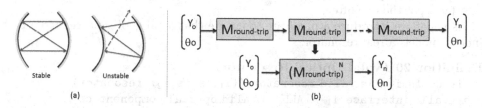

Fig. 5. (a) Types of Optical Resonators (b) Resonator Matrix After N Round-trips

between these interfaces. For example, consider a simple plane-mirror resonator as shown in Figure 4: let m_1 be the first mirror, m_2 the second one, and f the free space in between. Then the stability analysis involves the study of the ray as it goes through f, then reflects on m_2, then travels back through f, then reflects again on m_1, and starts over. So we have to consider the ray going through the "infinite" path $f, m_2, f, m_1, f, m_2, f, m_1, \ldots$, or, using regular expressions notations, $(f, m_2, f, m_1)^*$. Our purpose, regarding stability, is to ensure that this infinite ray remains inside the cavity. This is equivalent to consider that, for every n, the ray going through the path $(f, m_2, f, m_1)^n$ remains inside the cavity. This allows to reduce the study of an infinite path to an infinite set of finite paths.

Our formalization (which is inspired by the way optics engineers model optical systems), presented in Section 3.1, fixes the path of any considered ray. Since we want to consider an infinite set of finite-path rays, we should thus consider an infinite set of optical systems. This has been naturally achieved by optics engineers by "unfolding" the resonator as many times as needed, depending on the considered ray. For instance, consider again the above example of a plane-mirror resonator: if we want to observe a ray going back and forth only once through the cavity, then we should consider the optical system made of f, m_1, f, m_2; however, if we want to study the behavior of rays which make two round-trips through the cavity, then we consider a *new* optical system $f, m_1, f, m_2, f, m_1, f, m_2$; and similarly for more round-trips. This is the standard way optics engineers handle resonators and therefore is the one that we have chosen for our formalization, which we present now.

In our formalization, we want the user to provide only the minimum information so that HOL Light generates automatically the unfolded systems. Therefore, we do not define resonators as just optical systems but define a dedicated type for them: in their most general form, resonators are made of two reflecting interfaces and a list of components in between. We thus define the following type:

Definition 19 (Optical Resonator)
```
define_type "resonator = :interface × optical_component list ×
                         free_space × interface"
```

Note that the additional free space in the type definition is required because the `optical_component` type only contains one free space (the one before the interface, not the one after).

As usual, we introduce a predicate to ensure that a value of type `resonator` indeed models a real resonator:

Definition 20 (Valid Optical Resonator)
⊢ $\forall i_1$ cs fs i_2. is_valid_resonator ((i_1,cs,fs,i_2):resonator)⇔
   is_valid_interface i_1 ∧ ALL is_valid_optical_component cs ∧
   is_valid_free_space fs ∧ is_valid_interface i_1

We now present the formalization of the unfolding mentioned above. The first step in this process is to define a function `round_trip` which returns the list of components corresponding to one round-trip in the resonator:

Definition 21 (Round Trip)
⊢i_1 i_1 cs fs. round_trip ((i_1,cs,fs,i_2):resonator) =
   APPEND cs (CONS (fs,i_2,reflected)
   (let cs',fs_1 = optical_component_shift cs fs in
   REVERSE (CONS (fs$_1$,i$_1$,reflected) cs')))

where `APPEND` is a HOL Light library function which appends two lists, `REVERSE` reverses the order of elements of a list, and `optical_component_shift cs fs` shifts the free spaces of `cs` from right to left, introducing `fs` to the right; the leftmost free space which is "ejected" is also returned by the function. This manipulation is required because unfolding the resonator entails the reversal of the components for the return trip.

We can now define the unfolding of a resonator as follows:

Definition 22 (Unfold Resonator)
⊢ unfold_resonator ((i_1,cs,fs,i_2):resonator) N =
   list_pow (round_trip (i_1,cs,fs,i_2)) N,(head_index (cs,fs),0)

where `list_pow l n` concatenates `n` copies of the list `l`. The argument N represents the number of times we want to unfold the resonator. Note that the output type is `optical_system`, therefore all the previous predicates and theorems can be used on an unfolded resonator.

We can now define formally the notion of stability. For an optical resonator to be stable, the distance of the ray from the optical axis and its orientation should remain bounded whatever is the value of N. This is formalized as follows:

Definition 23 (Resonator Stability)
⊢ ∀res. is_stable_resonator res ⇔ (∀r. ∃y θ. ∀N.
  is_valid_ray_in_system r (unfold_resonator res N) ⟹
  (let yₙ,θₙ = last_single_ray r in abs(yₙ) ≤ y ∧ abs(θₙ) < θ))

Proving that a resonator satisfies the abstract condition of Definition 23 does not seem trivial at first. However, if the determinant of a resonator matrix M is 1 (which is the case in practice), optics engineers have known for a long time that having $-1 < \frac{M_{11}+M_{22}}{2} < 1$ is sufficient to ensure that the stability condition holds. The obvious advantage of this criterion is that it is immediate to check. This can actually be proved by using Sylvester's Theorem [26], which states that for a matrix $M = \begin{bmatrix} A & B \\ C & D \end{bmatrix}$ such that $\mid M \mid = 1$ and $-1 < \frac{A+D}{2} < 1$, the following holds:

$$\begin{bmatrix} A & B \\ C & D \end{bmatrix}^N = \frac{1}{\sin(\theta)} \begin{bmatrix} A\sin[N\theta] - \sin[(N-1)\theta] & B\sin[N\theta] \\ C\sin[N\theta] & D\sin[N\theta] - \sin[(N-1)\theta] \end{bmatrix}$$

where $\theta = cos^{-1}[\frac{A+D}{2}]$. This theorem allows to prove that stability holds under the considered assumptions: indeed, N only occurs under a sine in the resulting matrix; since the sine itself is comprised between -1 and 1, it follows that the components of the matrix are obviously bounded, hence the stability. We formalize Sylvester's theorem as follows:

Theorem 13 (Sylvester's Theorem)
⊢ ∀N A B C D. $\begin{vmatrix} A & B \\ C & D \end{vmatrix}$ = 1 ∧ $-1 < \frac{A+D}{2}$ ∧ $\frac{A+D}{2} < 1$ ⟹
  let $\theta = acs(\frac{(A+D)}{2})$ in
  $\begin{bmatrix} A & B \\ C & D \end{bmatrix}^N = \frac{1}{\sin(\theta)} \begin{bmatrix} A*\sin[N\theta] - \sin[(N-1)\theta] & B*\sin[N\theta] \\ C*\sin[N\theta] & D*\sin[N\theta] - \sin[(N-1)\theta] \end{bmatrix}$

We prove Theorem 13 by induction on N and using the fundamental properties of trigonometric functions, matrices and determinants. This allows to derive now the generalized stability theorem for any resonator as follows:

Theorem 14 (Stability Theorem)
⊢ ∀res. is_valid_resonator res ∧
  (∀N. let M = system_composition (unfold_resonator res 1) in
  det M = 1 ∧ $-1 < \frac{M_{1,1}+M_{2,2}}{2}$ ∧ $\frac{M_{1,1}+M_{2,2}}{2} < 1$) ⟹
  is_stable_resonator res

where $M_{i,j}$ represents the element at column i and row j of the matrix. The formal verification of Theorem 14 requires the definition of stability (Definition 23) and Sylvester's theorem along with the following important lemma:

Lemma 1 (Resonator Matrix)
⊢ ∀n res. system_composition (unfold_resonator res N)=
  system_composition (unfold_resonator res 1) mat_pow N

where `mat_pow` is the matrix power function. Note that it is an infix operator. This intuitive lemma formalizes the relation between the unfolding of a resonator and the corresponding ray-transfer matrix.

It is also important to note that our stability theorem is quite general and can be used to verify the stability of almost all kinds of optical resonators.

4   Applications

In this section, we present the stability of two widely used optical resonators: a Fabry Pérot resonator and a Z-shaped resonator.

4.1   Fabry Pérot Resonator

Nowadays, optical systems are becoming more and more popular due to their huge application potential. In order to bring this technology to the market, a lot of research has been done towards the integration of low cost, low power and portable building blocks in optical systems. One of the most important such building blocks is the Fabry Pérot (FP) resonator [19]. Originally, this resonator was used as a high resolution interferometer in astrophysical applications. Recently, the Fabry Pérot resonator has been realized as a microelectromechanical (MEMS) tuned optical filter for applications in reconfigurable Wavelength Division Multiplexing [18]. The other important applications are in the measurement of refractive index of cancer cells [24] and optical bio-sensing devices [2]. As a direct application of the

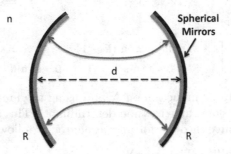

Fig. 6. Fabry Pérot resonator

framework developed in the previous sections, we present the stability analysis of the Fabry Pérot (FP) resonator with spherical mirrors as shown in Figure 6. This architecture is composed of two spherical mirrors with radius of curvature R separated by a distance d and refractive index n. We formally model this resonator as follows:

Definition 24 (FP Resonator)
⊢ ∀R d n. (fp_resonator R d n :resonator) =
        (spherical R,[],(n,d),spherical R)

where [] represents an empty list of components because the given structure has no component between spherical interfaces but only a free space (n,d). Next, we verify that the FP resonator is indeed a valid resonator as follows:

Theorem 15 (Valid FP resonator)
⊢ ∀R d n. R ≠ 0 ∧ 0 ≤ d ∧ 0 < n ⟹
  is_valid_resonator (fp_resonator R d n)

Finally, we formally verify the stability of the FP resonator as follows:.

Theorem 16 (Stability of FP Resonator)
⊢ ∀R d n. R ≠ 0 ∧ 0 < n ∧ 0 < $\frac{d}{2}$ ∧ $\frac{d}{2}$ < 2 ⟹
  is_stable_resonator (fp_resonator R d n)

The first two assumptions just ensure the validity of the model description. The two following ones provide the intended stability criteria. The formal verification of the above theorem requires Theorem 14 along with some fundamental properties of the matrices and arithmetic reasoning.

4.2 Z-Shaped Resonator

The Z-shaped resonator consists of two plane mirrors and two spherical mirrors as shown in Figure 7. It is widely used in many optical and laser systems including optical bandpass filters and all-optical timing recovery circuits [3]. We formally

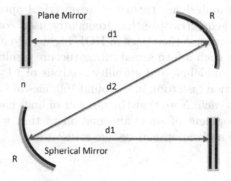

Fig. 7. Z-shaped resonator

model this resonator as follows:

Definition 25 (Z Resonator)
⊢ ∀R d_1 d_2 n. (z_resonator R d_1 d_2 n :resonator) = (plane,
  [(n,d_1),spherical R,reflected; (n,d_2),spherical R,reflected],
  (n,d_1),plane)

Here, we have a list of optical components and a free space between two plane mirrors. Again, we check the validity of the Z-shaped resonator as follows:

Theorem 17 (Valid Z-Shaped Resonator)
$\vdash \forall R\ d_1\ d_2\ n.\ R \neq 0 \wedge 0 \leq d_1 \wedge 0 \leq d_2 \wedge 0 < n \implies$
  is_valid_resonator (z_resonator R d_1 d_2 n)

Finally, we formally verify the stability of the FP resonator:

Theorem 18 (Stability of Z-Shaped Resonator)
$\vdash \forall R\ d_1\ d_2\ n.\ R \neq 0 \wedge 0 < d_1 \wedge 0 < n \wedge \frac{(2*d_1+d_2)^2}{2*d_1} < d_2 \wedge$
  $\frac{2*d_1*d_2}{2*d_1+d_2} < R \implies$ is_stable_resonator (z_resonator R d_1 d_2 n)

The first three assumptions just ensure the validity of the model description. The two following ones provide the intended stability criteria. The formal verification of the above theorem requires Theorem 14 along with some fundamental properties of the matrices and arithmetic reasoning.

4.3 Discussion

The formal stability analysis of the FP and Z-shaped resonators demonstrates the effectiveness of the proposed theorem proving based approach to reason about geometrical optics. Due to the formal nature of the model and inherent soundness of higher-order logic theorem proving, we have been able to formalize some foundations of geometrical optics along with the verification of some useful theorems about optical components and systems with an unrivaled accuracy. This improved accuracy comes at the cost of the time and effort spent, while formalizing the underlying theory of geometrical optics and resonators. The formalization of geometrical optics, frequently used components and resonators stability took around 1500 lines of HOL Light code and 250 man-hours. But the availability of such a formalized infrastructure significantly reduces the time required for the modeling and stability analysis of FP and Z-shaped resonators as the verification task took just around 100 lines of HOL Light code and a couple of man-hours each. Note that the number of lines has been significantly reduced by the development of some automation tactics, which automatically verifies the validity of a given optical system structure.

5 Conclusion

In this paper, we report a novel application of formal methods in analyzing optical and laser systems which is based on geometrical optics. We provided a brief introduction of the current state-of-the-art and highlighted their limitations. Next, we presented an overview of geometrical optics followed by highlights of our higher-order logic formalization. We also presented the formalization of frequently used optical components like thin lens, thick lens and plane parallel plate. In order to show the practical effectiveness of our formalization, we presented the stability analysis of two widely used optical resonators (i.e., Fabry Pérot resonator and Z-shaped resonator).

Our plan is to extend this work in order to obtain an extensive library of verified optical components, along with their ray-transfer matrices, which would allow a practical use of our formalization in industry. In addition, we plan to formally take into account the paraxial approximation using asymptotic notations [1]. We also intend to improve the traditional stability analysis method by handling infinite paths of rays (as described in Section 3.3) by working directly with all the possible paths of a ray, and thus avoiding the use of unfolding. In particular, this requires a more general treatment of optical interfaces without explicitly mentioning their behavior, i.e., transmitted or reflected. This is a very interesting direction of research since it would even go beyond what optics engineers currently do. Our long term goal is to package our HOL Light formalization in a GUI, so that it can be used by the non-formal methods community in industry for the analysis of practical resonators, and in academia for teaching and research purposes.

References

1. Avigad, J., Donnelly, K.: Formalizing O Notation in Isabelle/HOL. In: Basin, D., Rusinowitch, M. (eds.) IJCAR 2004. LNCS (LNAI), vol. 3097, pp. 357–371. Springer, Heidelberg (2004)
2. Baaske, M., Vollmer, F.: Optical Resonator Biosensors: Molecular Diagnostic and Nanoparticle Detection on an Integrated Platform. ChemPhysChem 13(2), 427–436 (2012)
3. Binh, L.N., Ngo, N.Q., Luk, S.F.: Graphical Representation and Analysis of the Z-shaped Double-Coupler Optical Resonator. Journal of Lightwave Technology 11(11), 1782–1792 (1993)
4. Cheng, Q., Cui, T.J., Zhang, C.: Waves in Planar Waveguide Containing Chiral Nihility Metamaterial. Optics and Communication 274, 317–321 (2007)
5. Gather, M.C., Yun, S.H.: Single-cell Biological Lasers. Nature Photonics 5(7), 406–410 (2011)
6. Gordon, M.J.C., Melham, T.F.: Introduction to HOL: A Theorem Proving Environment for Higher-Order Logic. Cambridge University Press (1993)
7. Harrison, J.: Theorem Proving with the Real Numbers. Springer (1998)
8. Harrison, J.: A HOL Theory of Euclidean Space. In: Hurd, J., Melham, T. (eds.) TPHOLs 2005. LNCS, vol. 3603, pp. 114–129. Springer, Heidelberg (2005)
9. Harrison, J.: Formalizing Basic Complex Analysis. In: From Insight to Proof: Festschrift in Honour of Andrzej Trybulec. Studies in Logic, Grammar and Rhetoric, vol. 10(23), pp. 151–165. University of Białystok (2007)
10. Hasan, O., Afshar, S.K., Tahar, S.: Formal Analysis of Optical Waveguides in HOL. In: Berghofer, S., Nipkow, T., Urban, C., Wenzel, M. (eds.) TPHOLs 2009. LNCS, vol. 5674, pp. 228–243. Springer, Heidelberg (2009)
11. Kogelnik, H., Li, T.: Laser Beams and Resonators. Appl. Opt. 5(10), 1550–1567 (1966)
12. LASCAD (2013), http://www.las-cad.com/
13. Malak, M., Pavy, N., Marty, F., Peter, Y., Liu, A.Q., Bourouina, T.: Stable, High-Q Fabry-Perot Resonators with Long Cavity Based on Curved, All-Silicon, High Reflectance Mirrors. In: IEEE 24th International Conference on Micro Electro Mechanical Systems (MEMS), pp. 720–723 (2011)

14. Mookherjea, S.: Analysis of Optical Pulse Propagation with Two-by-Two (ABCD) Matrices. Physical Review E 64(016611), 1–10 (2001)
15. Nakazawa, M., Kubota, H., Sahara, A., Tamura, K.: Time-domain ABCD Matrix Formalism for Laser Mode-Locking and Optical Pulse Transmission. IEEE Journal of Quantum Electronics 34(7), 1075–1081 (1998)
16. Naqvi, A.: Comments on Waves in Planar Waveguide Containing Chiral Nihility Metamaterial. Optics and Communication 284, 215–216 (2011)
17. OpticaSoftware (2013), http://www.opticasoftware.com/
18. Saadany, B., Malak, M., Kubota, M., Marty, F.M., Mita, Y., Khalil, D., Bourouina, T.: Free-Space Tunable and Drop Optical Filters Using Vertical Bragg Mirrors on Silicon. IEEE Journal of Selected Topics in Quantum Electronics 12(6), 1480–1488 (2006)
19. Saleh, B.E.A., Teich, M.C.: Fundamentals of Photonics. John Wiley & Sons, Inc. (1991)
20. Siddique, U., Aravantinos, V.: On the Formal Analysis of Geometrical Optics in HOL - HOL Light Script (2013),
    http://hvg.ece.concordia.ca/code/hol-light/goptics/
21. Siddique, U., Aravantinos, V., Tahar, S.: Higher-Order Logic Formalization of Geometrical Optics. In: International Workshop on Automated Deduction in Geometry, pp. 185–196. Informatics Research Report, School of Informatics, University of Edinburgh, UK (2012)
22. Siddique, U., Hasan, O.: Formal Analysis of Fractional Order Systems in HOL. In: Proceedings of the IEEE International Conference on Formal Methods in Computer-Aided Design (FMCAD), Austin, TX, USA, pp. 163–170 (2011)
23. Siegman, A.E.: Lasers, 1st edn. University Science Books (1986)
24. Song, W.Z., Zhang, X.M., Liu, A.Q., Lim, C.S., Yap, P.H., Hosseini, H.M.M.: Refractive Index Measurement of Single Living Cells Using On-Chip Fabry-Perot Cavity. Applied Physics Letters 89(20), 203901 (2006)
25. Su, B., Xue, J., Sun, L., Zhao, H., Pei, X.: Generalised ABCD Matrix Treatment for Laser Resonators and Beam Propagation. Optics & Laser Technology 43(7), 1318–1320 (2011)
26. Sylvester, J.J.: The Collected Mathematical Papers of James Joseph Sylvester, vol. 4. Cambridge U. Press (1912)
27. Loesel, F.H., Kurtz, R.M., Horvath, C., Bille, J.F., Juhasz, T., Djotyan, G., Mourou, G.: Applications of Femtosecond Lasers in Corneal Surgery. Laser Physics 10(2), 495–500 (2011)
28. Wilson, W.C., Atkinson, G.M.: MOEMS Modeling Using the Geometrical Matrix Toolbox. Technical report, NASA, Langley Research Center (2005)

Preprocessing of the Axiomatic System
for More Efficient Automated Proving
and Shorter Proofs*

Sana Stojanović

Faculty of Mathematics, University of Belgrade
Studentski trg 16, 11000 Belgrade, Serbia
sana@matf.bg.ac.rs

Abstract. One of the main differences between pen-and-paper proofs
and computer proofs is in the number of simple facts derived. In automated proof generation, the number of simple facts can be large. We are
addressing this problem by preprocessing of the axiomatic system that
should enable reduction in the number of simple and redundant facts
to some extent. We implemented two types of preprocessing techniques,
one concerning symmetric predicates, and another restricting introduction of witnesses during proof search. Both techniques were used within
a coherent logic prover ArgoCLP. Evaluations performed on geometrical
domain show that use of these techniques makes automated process more
efficient and generated proofs often significantly shorter.

Keywords: Predicate symmetry, Axiom reformulation, Coherent logic,
Automated and formal theorem proving.

1   Introduction

One of common challenges in automated and interactive theorem proving is cop-
ing with proving "simple facts". In traditional pen-and-paper theorem proving,
"simple facts" – trivial facts that don't need deep proofs and theory-specific arguments – are typically assumed or neglected. However, in automated theorem
proving "simple facts" can significantly increase the search space, while in interactive theorem proving they can make a significant burden in following the main
line of the proof. Because of this, there is a number of methods and techniques
for dealing with simple facts both in automated and interactive theorem proving.
In this paper, we will address this issues in the context of automated generation
of formal (machine verifiable) but also readable proofs.

We address two sorts of simple facts – those that rely on symmetry properties
of certain predicates and those that rely on specific reformulations of certain
axioms or lemmas. Symmetries and properties of symmetrical predicates have

---

* This work has been partly supported by the grant 174021 of the Ministry of Science
  of Serbia and by the SNF SCOPES grant IZ73Z0_127979/1.

T. Ida and J. Fleuriot (Eds.): ADG 2012, LNAI 7993, pp. 181–192, 2013.

been widely studied and used in automated reasoning [5,6,7], but in this paper we will consider one variant sufficient for our purposes. In cases when the prover introduces witnesses during the proof search, a large number of (potentially unnecessary) witnesses can lead to larger search space and redundant steps.

As an answer to those problems we propose two simple preprocessing techniques. We built these two techniques into ArgoCLP [12] – a prover based on coherent logic and forward reasoning that generates machine verifiable proofs in Isar [11] and proofs in natural language. In the evaluation, we are focusing on Euclidean geometry. The evaluation shows that application of these two techniques improves the efficiency of the prover and generate proofs which are often significantly shorter. Moreover, the readable proofs that ArgoCLP generates become more similar to the proofs that can be found in mathematics textbooks.

This paper is organized as follows: in Section 2 we give background information on coherent logic and on prover ArgoCLP; in Section 3 we describe the importance of symmetric predicates, their detection and use; in Section 4 we describe a technique dealing with axioms that introduce several witnesses; in Section 5 we give an overview of our experiments; in Section 6 we give an overview of related work; and in Section 7 we give our conclusions and plans for the future work.

2   Coherent Logic and ArgoCLP Prover

Coherent Logic. Coherent logic (CL) was initially defined by Skolem and in recent years was popularized by Bezem [3]. It is a fragment of first-order logic, consisting of implicitly universally quantified formulae of the following form:

$$A_1(\boldsymbol{x}) \wedge \ldots \wedge A_n(\boldsymbol{x}) \Rightarrow \exists \boldsymbol{y_1} B_1(\boldsymbol{x}, \boldsymbol{y_1}) \vee \ldots \vee \exists \boldsymbol{y_m} B_m(\boldsymbol{x}, \boldsymbol{y_m}) \qquad (2.1)$$

where: $n \geq 0$, $m \geq 0$, $\boldsymbol{x}$ denotes a sequence of variables, A_i (for $1 \leq i \leq n$) denotes an atomic formula (involving some of the variables from $\boldsymbol{x}$), $\boldsymbol{y_j}$ denotes a sequence of variables, and B_j (for $1 \leq j \leq m$) denotes a conjunction of atomic formulae (involving some of the variables from $\boldsymbol{x}$ and $\boldsymbol{y_j}$). There are no function symbols with arity greater than 0. Function symbols of arity 0 are called *constants*. A *witness* is a new constant, not appearing in axioms used nor in the conjecture being proved. A *term* is a constant or a variable. An *atomic formula* is either $\perp$ or $p(t_1, \ldots, t_n)$ where p is a predicate symbol of arity n and t_i $(1 \leq i \leq n)$ are terms. An atomic formula over constants is called a *fact*. CL deals with the sets of facts — ground atomic expressions.

The reasoning in coherent logic is constructive and proof objects can easily be obtained, therefore CL is suitable for producing both readable and formal proofs. A large number of theories and theorems can be formulated directly in CL. There exists a linear translation from FOL to CL that preserves logical equivalence.

ArgoCLP Prover. ArgoCLP [12] is a generic theorem prover based on coherent logic that automatically produces formal proofs in Isar and readable proofs in

English that resemble proofs that can be found in mathematics textbooks. It can be used with any set of coherent axioms. The proof procedure is simple forward chaining with iterative deepening. Since negations are not supported in coherent logic, for every predicate symbol R, typically an additional symbol $\overline{R}$ is introduced (that stands for $\neg R$) and the following two axioms are added to the set of axioms (for every predicate R): $R(x) \wedge \overline{R}(x) \Rightarrow \perp$, and $R(x) \vee \overline{R}(x)$ (section A.1 of the appendix).

The following example shows a proof of one geometry theorem in natural language generated by the ArgoCLP prover.

Example 1.   Proof generated by the ArgoCLP prover.

Theorem: Assuming that $p \neq q$, and $q \neq r$, and the line p is incident to the plane α, and the line q is incident to the plane α, and the line r is incident to the plane α, and the lines p and q do not intersect, and the lines q and r do not intersect, and the point A is incident to the plane α, and the point A is incident to the line p, and the point A is incident to the line r, show that $p = r$.

Proof

Let us prove that $p = r$ by reductio ad absurdum.

  1. Assume that $p \neq r$.

  2. It holds that the point A is incident to the line q or the point A is not incident to the line q (by axiom of excluded middle).

  3. Assume that the point A is incident to the line q.

  4. From the facts that $p \neq q$, and the point A is incident to the line p, and the point A is incident to the line q, it holds that the lines p and q intersect (by axiom ax_D5).

  5. From the facts that the lines p and q intersect, and the lines p and q do not intersect we get a contradiction.

  Contradiction.

  6. Assume that the point A is not incident to the line q.

  7. From the facts that the lines p and q do not intersect, it holds that the lines q and p do not intersect (by axiom ax_nint_1_1_21).

  8. From the facts that the point A is not incident to the line q, and the point A is incident to the plane α, and the line q is incident to the plane α, and the point A is incident to the line p, and the line p is incident to the plane α, and the lines q and p do not intersect, and the point A is incident to the line r, and the line r is incident to the plane α, and the lines q and r do not intersect, it holds that $p = r$ (by axiom ax_E2).

  9. From the facts that $p = r$, and $p \neq r$ we get a contradiction.

  Contradiction.

Therefore, it holds that $p = r$.

This proves the conjecture.

Theorem proved in 9 steps and in 0.02 s.

3   Dealing with Symmetric Predicates

Dealing with symmetric predicates is standardly supported in many automated theorem provers, in different forms. In this paper we are attempting to automatically add support for symmetric predicates as preprocessing technique.

Definition 1. *(Symmetric predicate) An n-ary predicate R is symmetric (in all arguments) if the following (universally quantified) statement holds in the considered theory for every permutation σ:*

$$R(x_1, \ldots, x_n) \Leftrightarrow R(x_{\sigma(1)}, \ldots, x_{\sigma(n)})$$

Theorem 1. *An n-ary predicate R is symmetric if and only if the following two (universally quantified) statements hold:*

$$R(x_1, x_2, x_3, \ldots, x_n) \Leftrightarrow R(x_2, x_1, x_3 \ldots, x_n) \tag{3.1}$$

$$R(x_1, x_2, x_3 \ldots, x_n) \Leftrightarrow R(x_2, x_3, \ldots, x_n, x_1) \tag{3.2}$$

In case that these two formulae are lemmas, formulae that express symmetry of all other permutations are lemmas as well. Those lemmas and their proofs are important for automated theorem proving and for generating formal proofs. The set of all lemmas that express symmetry of a predicate will be referred to as *the symmetry lemmas.*

Preprocessing Phase. The statements (3.1) and (3.2) can be automatically generated from the set of predicates of the axiomatic system, and the prover can then automatically check whether these formulae are lemmas of the considered theory (since not all of them will be lemmas, a time restriction must be set). For symmetric predicates we can automatically generate a set of all symmetry lemmas which can then be used in the automatic generation of formal proofs in a manner described bellow.

Automated Proving Phase. In order to efficiently manage symmetric predicates, all permutations of arguments of ground atomic formula will be represented by a single permutation — a sorted one (for instance, using lexicographic ordering). For example, for the predicate of collinearity and constants A, B, C, ground atomic formulae $col(B, A, C)$ and $col(A, C, B)$ will both be represented with the ground atom $col(A, B, C)$, and are considered the same during proof search.

Object Level Proof Construction Phase. During proof generation these steps are complemented with the following two lemmas:

$col(B, A, C) \Rightarrow col(A, B, C)$
$col(A, B, C) \Rightarrow col(C, A, B)$

Assuming that these lemmas were already proven by the prover, a complete (formal) proof can be generated.

4   Dealing with Axiom Reformulations

Consider a (coherent) axiom of the following form:

$$A_1(x) \wedge \ldots \wedge A_n(x) \Rightarrow \exists y B(x, y) \qquad (4.1)$$

where $n \geq 0$, $y = \{y_1, y_2, \ldots, y_k\}$ and $k \geq 2$. During the proof search, in the context of forward chaining, this axiom *introduces k witnesses*:

$$A_1(x) \wedge \ldots \wedge A_n(x) \Rightarrow \exists y_1 \exists y_2 \ldots \exists y_k B(x, y_1, \ldots, y_k)$$

In case when $k = 1$ the coherent logic prover does not apply axiom of the form $A_1(x) \wedge \ldots \wedge A_n(x) \Rightarrow \exists y_1 B(x, y_1)$ if there is a such that $B(x, a)$ holds, but in the case of k existential quantifiers $(k > 1)$ it checks for all k witnesses. I.e., this axiom will not be applied if all of the witnesses instantiating $y_1, \ldots, y_k$ already exist, but if at least one of them is missing, the axiom will introduce k new witnesses. Our goal is to introduce less than k witnesses in case when some of the witnesses that satisfy the axiom already exists.

Let the formula B be a conjunction of atoms, such that it can be represented as $B = B_1 \wedge B_2$, where B_1 is a conjunction of all atoms that have *only* variables from x and y_1 (if such atoms do not exist, B_1 is $\top$), and B_2 is a non-empty conjunction of all other atoms from B:

$$A_1(x) \wedge \ldots \wedge A_n(x) \Rightarrow \exists y_1 \ldots \exists y_k (B_1(x, y_1) \wedge B_2(x, y_1, \ldots, y_k))$$

Consider the transformed version of the previous statement:

$$A_1(x) \wedge \ldots \wedge A_n(x) \wedge B_1(x, y_1) \Rightarrow \exists y_2 \ldots \exists y_k B_2(x, y_1, \ldots, y_k) \qquad (4.2)$$

This statement introduces less witnesses that the original one. In a general case, such statement is not a consequent of axioms and is not provable within a given theory. In case that it is provable, it will be used as a lemma but with higher priority than the original axiom.

Preprocessing Phase. Axioms of the form 4.1 can be automatically recognized (from the set of axioms) and a transformed statement can be generated for each of them. The prover can check if a transformed statement is provable (since not all generated statements will be provable, time restriction must be set). If the transformed statement is provable, the transformation can be applied for the new formula. This process is iterated while there can be found non empty formula B_2.

Automated Proving Phase. The generated lemmas are used as axioms, but during a proof search the prover gives higher priority to those lemmas over the original axiom.

Example 2. Considering the proofs of theorems based on Hilbert's axioms, we notice that certain axioms are rarely applied in their original form. For example, the axiom *I3: On every line there lie two different points*, is more often applied in the following manner *(I3a): If there is a point A on the line p, then there is a point B which is different from A and lies on p.* This statement does not correspond to a verbatim application of that axiom. We should actually introduce two new points B and C, such that B and C are different and that both of them lie on p. Nevertheless, this manner of application of axiom $I3$ is standardly used in mathematical proofs. The statement $I3a$ can actually be proven as a theorem, and that would justify using it.

A problem that occurs with this approach is that not all generated statements will be theorems. Let us consider the following example:

Example 3. Axiom *I8: There exist three non-collinear points,* will generate the following two statements:

1. *Given a point A, there exist points B and C such that A, B and C are non-collinear.*
2. *Given points A and B, there exists a point C such that A, B and C are non-collinear.*

The first statement is a theorem, but the second statement is not a theorem. It lacks an additional condition in its premises, i.e., points A and B must be different. In such case, user may be prompted to try to assist and to add missing premises.

5   Applications in Euclidean Geometry

In this section we discuss both the efficiency of the presented preprocessing techniques, and the effects that they have on the power of the prover. Both techniques were implemented in the prover ArgoCLP. The tool that implements preprocessing techniques is separated from the prover itself. That way, the generation of auxiliary theorems is performed only once for one axiomatic system and need not be performed every time when proving a theorem. All experiments were performed on AMD Opteron 2GHz with 96GB RAM[1]. The system was applied on a Hilbert style axiomatic system (only axioms of the first group of Hilbert's axioms were used). Most of Hilbert's axioms and theorems are directly expressible in coherent logic[2].

Axiomatic System. Our axiomatic system is based on Hilbert's axiomatic system [8]. Since ArgoCLP works with coherent logic, some Hilbert's axioms had to be transformed in coherent logic form. The main transformation is elimination of negation. As discussed in Section 2, for each predicate[3] R new predicate $\overline{R}$

---

[1] All materials can be found in http://www.matf.bg.ac.rs/~sana/system.zip

[2] Avigad, Dean, and Mumma [1] also noticed that a strong syntactic restriction on formulae, similar to the coherent logic, is adequate to representing the axioms and theorems of Euclid's plane geometry.

[3] Incidence of point with a line, incidence of point with a plane, incidence of line with a plane; intersection of two lines, intersection of two planes; collinearity of three points, coplanarity of four points, relation between for three points, congruence between pairs of points.

is added and the following axiom (definition) is added to the axiomatic system: $R \wedge \overline{R} \Rightarrow \bot$. For some predicates R it is easy to define these new predicates $\overline{R}$ explicitly[4]. Alternatively, the following axiom can be used (which is less preferred, because of cases split it introduces): $R \vee \overline{R}$. For predicates that are defined, this formula can be proven as a theorem[5]. Those definitions are mainly trivial and they are presented in appendix A.1.

Automated Detection of Symmetric Predicates. Symmetric property of predicates R and $\overline{R}$ are proven separately because symmetry lemmas for both of those predicates are used in completion of proofs.

Only predicates whose arguments are all of the same type are processed (positive and negative form of intersection of lines, intersection of planes, between, collinearity, coplanarity, and congruence). A total of 20 statements of the form (3.1) and (3.2) was generated (for predicates of arity two, these statements are identical). Only those predicates for which both statements were proved can be used as symmetric in proving geometry theorems.

In the set of the listed predicates, the following eight are symmetric: positive and negative form of intersection of lines, intersection of planes, collinearity and coplanarity. ArgoCLP succeeded in proving that seven out of these eight predicates are symmetric (all but $\overline{coplanarity}$) with the average execution time 3.6 seconds and the average number of steps 170. For predicates that are proven to be symmetric, all lemmas that express symmetry of the predicate by permutations not covered by (3.1) and (3.2) are generated and again ArgoCLP was used to generate their proofs. There are 40 such lemmas in total (with proving time under 2 seconds).

Automated Reformulation of Axioms. The set of axioms that are automatically recognized as axioms which introduce more than one witness is:

I3a There exist at least two different points on a line.
I3b There exist at least three points that are non-collinear.
I8 There exist at least four points which are non-coplanar.[6]

Statements that are automatically generated from this set, in the manner described in section 4, are:

I3a1 $(\forall p : Line)(\forall A : Point)A \in p \Rightarrow (\exists B : Point)(A \neq B \wedge B \in p)$
I3b1 $(\forall A : Point)(\exists B : Point)(\exists C : Point)\neg col(A, B, C)$
I3b2 $(\forall A : Point)(\forall B : Point)(\exists C : Point)\neg col(A, B, C)$

---

[4] $\overline{col}(A, B, C) \leftrightarrow (\exists p)(A \in p \wedge B \in p \wedge C \notin p)$.

[5] For example, if $A \in p \vee \overline{A \in p}$ holds, then it is trivial to prove $col(A, P, Q) \vee \overline{col}(A, P, Q)$ (with appropriately defined predicate $\overline{col}$).

[6] These are not original formulations of the axioms. Axioms had to be slightly modified in order to express them in coherent logic. Original formulation of these axioms is: [I3a] There exist at least two points on a line, [I3b] There exist at least three points that do not lie on a line, [I8] There exist at least four points which do not lie in a plane.

I8a $(\forall A : Point)(\exists B : Point)(\exists C : Point)(\exists D : Point)\neg cop(A, B, C, D)$
I8b $(\forall A : Point)(\forall B : Point)(\exists C : Point)(\exists D : Point)\neg cop(A, B, C, D)$
I8c $(\forall A : Point)(\forall B : Point)(\forall C : Point)(\exists D : Point)\neg cop(A, B, C, D)$

Two of these statements were proven (theorems $I3a1$ and $I3b1$), and formal proofs in Isar were generated. Statements $I3b2$ and $I8b$ lack the condition $A \neq B$, and statement $I8c$ lacks the condition $\neg col(A, B, C)$ in order to be theorems. Total time of this preprocessing phase is 20 minutes (when time limit is set to 5 minutes).

Improvement in the Performance of ArgoCLP. The described techniques were tested on the prover ArgoCLP with small benchmark set of 24 theorems[7] that were all proved with ArgoCLP. The median and average proving time were 2 minutes and 23 minutes (due to four hard lemmas), while the average number of steps was 1374.

1. *Effect of Symmetric Predicates.* For this evaluation we used only the automatically proven symmetry theorems (that express symmetry for 7 out of 8 symmetric predicates). With this technique average proving time was reduced by 56% and the average number of steps by 83% (average proving time was 10 minutes and average number of steps 230).
2. *Exploitation of Axiom Reformulation.* For this evaluation we used only the reformulated versions of axioms that were proven automatically (two lemmas). We compare the performance of the prover using both techniques to its performance when using only the technique for dealing with symmetries. The average proving time was reduced by 15% and the average number of steps by 37% (average proving time was 8.5 minutes and average number of steps 143.).
3. *The total improvement* when using both techniques compared to the original prover is 63% in the average proving time and 89% in the average number of steps.

6   Related Work

Techniques for handling symmetric predicates are widely used in automated reasoning and constraint programming. The symmetry of problem instances (propositional formulae, CSPs, etc.), are intensively exploited in theorem proving. For instance, Arai and Masukawa [2] designed a ground theorem prover Godzila that quickly finds symmetries in combinatorial problems. Cadoli and Mancini [4,9] considered specifications of constraint problems as logical formulae, and used automated theorem prover to determine existence of symmetries and functional dependencies.

---

[7] For newly introduced predicates that are defined with new axioms, excluded middle rule is a theorem and is a part of this set.

Dealing with symmetries in geometry dates back to 1959 and Gelernter [7]. He was dealing with the problem of efficient machine manipulation of formal systems in which the predicates display a high degree of symmetry, and successfully eliminated symmetry-redundant goals.

Chou, Gao, and Zhang developed a deductive database method [6] that can be used to prove or discover nontrivial geometry theorems. They noticed that most geometric predicates (such as collinearity) satisfy properties such as transitivity and symmetry, which leads to a large database and repetitive representation of information. They used equivalence classes and canonical forms (in a way that is similar to ours) to represent facts in the database and reduced size of the database by a factor of 100.

Caferra, Peltier, and Puitg [5] used a similar approach in geometry theorem proving in order to reduce the number of facts conveying the same information. They incorporated human techniques in the prover which increased its power and made user interaction more natural. They encoded equational theories such as commutativity and circularity in the unification algorithm. Only the minimal representative of equivalence class (determined by commutativity and circularity) is stored into the database which results in a significant speed up.

Stojanović, Pavlović, and Janičić [12] developed a prover ArgoCLP but analyzed symmetries in a Hilbert style axiomatic system by hand. The information on all symmetric predicates (all eight predicates, positive and negative form of four symmetric predicates) was given to the prover as a new set of lemmas. Also, reformulation of axioms was done manually and all reformulated formulae (six) were added to the set of lemmas (after human intervention and inclusion of the missing relations). In this paper we showed that to some extent this process can be done automatically, without human intervention.

Meikle and Fleuriot [10] developed a semi-automated approach[8] to geometry theorem proving. They automated in Isabelle/HOL some of the geometric reasoning (symmetry of collinearity) that is often required in verification tasks by extending Isabelle's simplifier and classical reasoner. They noticed that some of those theorems alone could generate infinite loops and that special attention is needed. The potential problem with symmetry theorems and infinite loops are not present with our approach either. The theorems that can generate an infinite loop will not be used by our prover during the proof search, but only to complete the proofs in places where the knowledge about symmetry is needed. In contrast to their approach where proofs of theorems that express symmetry of collinearity were proved by hand, in our approach, all symmetry theorems that are used were proven automatically.

We are not aware of prior work related to axiom reformulation.

---

[8] Semi-automated theorem proving is using automated techniques that are implemented within interactive theorem proving assistant. During interactive process, the user has the option of using the help of tool for automation if it manages to derive some useful goals.

7 Conclusions and Future Work

We proposed two techniques for modification of an axiomatic system that improve efficiency of forward chaining automated theorem provers. These techniques were implemented and tested within ArgoCLP prover and the evaluation was performed on the theorems of Euclidean geometry using Hilbert like axiomatic system.

Automatic detection of symmetries managed to detect 7 out of 8 symmetric predicates. The automated reformulation of axioms managed to detect 2 out of 6 useful lemmas. We showed that the exploitation of this knowledge during the work of the prover can give better performance compared to the original prover. The execution time was shortened by 63%. In some cases the proposed techniques do not provide any speedup, but they never result in decreased efficiency.

With these simple techniques the generated proofs are shorter, simpler, and closer to proofs from mathematics textbooks. These techniques can be useful, for instance, for checking equivalence of different axiomatic systems (Hilbert's axiomatic system has several different interpretation).

Apart from predicates that are symmetric on all arguments, there are predicates which are symmetric only in some arguments (*between, congruence*). They are not discussed in this paper but can be treated in the same manner.

Most of the statements generated by reformulation of axioms in our experiments are not provable because they are missing additional condition in premises (4 out of 6 generated statements are not provable). One way of dealing with this problem is user assistance and modification of those statements manually. In the future, we plan to devise a heuristic for automatic discovery of additional conditions.

References

1. Avigad, J., Dean, E., Mumma, J.: A formal system for Euclid's Elements. The Review of Symbolic Logic (2009)
2. Arai, N.H., Masukawa, R.: How to Find Symmetries Hidden in Combinatorial Problems. In: Proceedings of the Eighth Symposium on the Integration of Symbolic Computation and Mechanized Reasoning (2000)
3. Bezem, M., Coquand, T.: Automating coherent logic. In: Sutcliffe, G., Voronkov, A. (eds.) LPAR 2005. LNCS (LNAI), vol. 3835, pp. 246–260. Springer, Heidelberg (2005)
4. Cadoli, M., Mancini, T.: Using a Theorem Prover for Reasoning on Constraint Problems. In: Bandini, S., Manzoni, S. (eds.) AI*IA 2005. LNCS (LNAI), vol. 3673, pp. 38–49. Springer, Heidelberg (2005)
5. Caferra, R., Peltier, N., Puitg, F.: Emphasizing Human Techniques in Automated Geometry Theorem Proving A Practical Realization. In: Richter-Gebert, J., Wang, D. (eds.) ADG 2000. LNCS (LNAI), vol. 2061, pp. 268–305. Springer, Heidelberg (2001)
6. Chou, S.-C., Gao, X.-S., Zhang, J.-Z.: A Deductive Database Approach to Automated Geometry Theorem Proving and Discovering. Journal of Automated Reasoning (2000)

7. Gelernter, H.: A Note on Syntatic Symmetry and the Manipulation of Formal Systems by Machine. Information and Control (1959)
8. Hilbert, D.: Grundlagen der Geometrie, Leipzig (1899)
9. Mancini, T., Cadoli, M.: Detecting and Breaking Symmetries by Reasoning on Problem Specifications. In: Zucker, J.-D., Saitta, L. (eds.) SARA 2005. LNCS (LNAI), vol. 3607, pp. 165–181. Springer, Heidelberg (2005)
10. Meikle, L., Fleuriot, J.: Automation for Geometry in Isabelle/HOL. In: Proceedings of PAAR, FLOC 2010 (2010)
11. Nipkow, T., Paulson, L.C., Wenzel, M.: Isabelle/HOL. LNCS, vol. 2283. Springer, Heidelberg (2002)
12. Stojanović, S., Pavlović, V., Janičić, P.: A Coherent Logic Based Geometry Theorem Prover Capable of Producing Formal and Readable Proofs. In: Schreck, P., Narboux, J., Richter-Gebert, J. (eds.) ADG 2010. LNCS (LNAI), vol. 6877, pp. 201–220. Springer, Heidelberg (2011)

A   Hilbert's Axiomatic System I Group of Axioms

For readability, instead of writing $\overline{A \in p}$ we will write $A \notin p$, and instead of writing $\overline{col}(A, B, C)$ we will write $\neg col(A, B, C)$.

Universal quantifiers and types are omitted in the formule for readability. Capital letters are used for points, small letters are used for lines, and Greek letters are used for planes.

I1 $A \neq B \Rightarrow (\exists p : line)(A \in p \wedge B \in p)$
I2 $A \neq B \wedge A \in p \wedge B \in p \wedge A \in q \wedge B \in q \Rightarrow p = q$
I3a $(\exists A : point)(\exists B : point)(A \neq B \wedge A \in p \wedge B \in p)$
I3b $(\exists A : point)(\exists B : point)(\exists C : point)\neg col(A, B, C)$
I4a $\neg col(A, B, C) \Rightarrow (\exists \alpha : plane)(A \in \alpha \wedge B \in \alpha \wedge C \in \alpha)$
I4b $(\exists A : point)A \in \alpha$
I5 $\neg col(A, B, C) \wedge A \in \alpha \wedge B \in \alpha \wedge C \in \alpha \wedge A \in \beta \wedge B \in \beta \wedge C \in \beta \Rightarrow \alpha = \beta$
I6 $A \neq B \wedge A \in p \wedge A \in \alpha \wedge B \in p \wedge B \in \alpha \Rightarrow p \in \alpha$
I7 $\alpha \neq \beta \wedge A \in \alpha \wedge A \in \beta \Rightarrow (\exists B : point)(A \neq B \wedge B \in \alpha \wedge B \in \beta)$
I8 $(\exists A : point)(\exists B : point)(\exists C : point)(\exists D : point)\neg cop(A, B, C, D)$

A.1   Additional Axioms

1. $A = B \vee A \neq B$
2. $p = q \vee p \neq q$
3. $\alpha = \beta \vee \alpha \neq \beta$
4. $A \in p \vee A \notin p$
5. $A \in \alpha \vee A \notin \alpha$
6. $A \in p \wedge B \in p \wedge C \in p \Rightarrow col(A, B, C)$
7. $col(A, B, C) \Rightarrow (\exists p : line)(A \in p \wedge B \in p \wedge C \in p)$
8. $A \neq B \wedge A \in p \wedge B \in p \wedge C \notin p \Rightarrow \neg col(A, B, C)$
9. $A \in \alpha \wedge B \in \alpha \wedge C \in \alpha \wedge D \in \alpha \Rightarrow cop(A, B, C, D)$
10. $cop(A, B, C, D) \Rightarrow (\exists \alpha : plane)(A \in \alpha \wedge B \in \alpha \wedge C \in \alpha \wedge D \in \alpha)$
11. $\neg col(A, B, C) \wedge A \in \alpha \wedge B \in \alpha \wedge C \in \alpha \wedge D \notin \alpha \Rightarrow \neg cop(A, B, C, D)$

12. $p \neq q \wedge A \in p \wedge A \in q \Rightarrow int(p,q)$
13. $int(p,q) \Rightarrow (\exists A : point)(A \in p \wedge A \in q \wedge p \neq q)$
14. $int(p,q) \vee \neg int(p,q)$
15. $\alpha \neq \beta \wedge A \in \alpha \wedge A \in \beta \Rightarrow int(\alpha,\beta)$
16. $int(\alpha,\beta) \Rightarrow (\exists A : point)(A \in \alpha \wedge A \in \beta \wedge \alpha \neq \beta)$
17. $int(\alpha,\beta) \vee \neg int(\alpha,\beta)$
18. $p \notin \alpha \wedge A \in p \wedge A \in \alpha \Rightarrow int(p,\alpha)$
19. $int(p,\alpha) \Rightarrow (\exists A : point)(A \in p \wedge A \in \alpha \wedge p \notin \alpha)$
20. $int(p,\alpha) \vee \neg int(p,\alpha)$
21. $p \in \alpha \wedge A \in p \Rightarrow A \in \alpha$
22. $A \in p \wedge A \notin \alpha \Rightarrow p \notin \alpha$

A.2   Theorems

1. $A \neq B \wedge col(A,B,C) \wedge col(A,B,D) \Rightarrow col(A,C,D)$
2. $col(A,B,C) \vee \neg col(A,B,C)$
3. $A \neq C \wedge A \in p \wedge B \notin p \wedge C \in p \Rightarrow \neg col(A,B,C)$
4. $A \neq C \wedge A \in p \wedge C \in p \wedge \neg col(A,B,C) \Rightarrow B \notin p$
5. $p \neq q \wedge \neg int(p,q) \wedge A \in p \Rightarrow A \notin q$
6. $\alpha \neq \beta \wedge \neg int(\alpha,\beta) \wedge A \in \alpha \Rightarrow A \notin \beta$
7. $p \notin \alpha \wedge A \in p \Rightarrow A \notin \alpha$
8. $\neg col(A,B,C) \Rightarrow (\exists \alpha : plane)(A \in \alpha \wedge B \in \alpha \wedge C \in \alpha)$
9. $cop(A,B,C,D) \wedge \neg col(A,B,C) \wedge A \in \alpha \wedge B \in \alpha \wedge C \in \alpha \Rightarrow D \in \alpha$
10. $p \in \alpha \vee p \notin \alpha$
11. $\neg col(A,B,C) \Rightarrow A \neq B \wedge A \neq C \wedge B \neq C$
12. $(\exists A : point)(\exists B : point)A \neq B$
13. $col(A,A,A)$
14. $(\exists A : point)A \notin p$
15. $p \in \alpha \wedge q \in \alpha \Rightarrow (\exists A : point)(A \in p \wedge A \in q) \vee \neg int(p,q)$
16. $p \neq q \wedge A \in p \wedge A \in q \wedge B \in p \wedge B \in q \Rightarrow A = B$
17. $\neg int(\alpha,\beta) \vee ((\exists p : line)p \in \alpha \wedge p \in \beta)$
18. $\alpha \neq \beta \wedge p \in \alpha \wedge p \in \beta \wedge A \in \alpha \wedge A \in \beta \Rightarrow A \in p$
19. $p \notin \alpha \Rightarrow (\exists A : point)(A \in p \wedge A \in \alpha) \vee \neg int(p,\alpha)$
20. $p \notin \alpha \wedge A \in p \wedge A \in \alpha \wedge B \in p \wedge B \in \alpha \Rightarrow A = B$
21. $A \notin p \Rightarrow (\exists \alpha : plane)(A \in \alpha \wedge p \in \alpha)$
22. $A \notin p \wedge p \in \alpha \wedge A \in \alpha \wedge p \in \beta \wedge A \in \beta \Rightarrow \alpha = \beta$
23. $p \neq q \wedge A \in p \wedge A \in q \Rightarrow (\exists \alpha : plane)(p \in \alpha \wedge q \in \alpha)$
24. $p \neq q \wedge A \in p \wedge A \in q \wedge p \in \alpha \wedge q \in \alpha \wedge p \in \beta \wedge q \in \beta \Rightarrow \alpha = \beta$

Author Index